STATIC CROSSTALK-NOISE ANALYSIS
For Deep Sub-Micron Digital Designs

STATIC CROSSTALK-NOISE ANALYSIS
For Deep Sub-Micron Digital Designs

by

Pinhong Chen
Cadence Design Systems, Inc.
Desmond A. Kirkpatrick
Intel Corporation
Kurt Keutzer
University of California, Berkeley

SPRINGER SCIENCE+BUSINESS MEDIA, LLC

Library of Congress Cataloging-in-Publication

Title: Static Crosstalk-Noise Analysis for Deep Sub-Micron Digital Designs
Author (s): Pinhong Chen, Desmond A. Kirkpatrick and Kurt Keutzer
ISBN 978-1-4757-7949-3   ISBN 978-1-4020-8092-0 (eBook)
DOI 10.1007/978-1-4020-8092-0

Contents

Contents vii

List of Figures

List of Tables

Preface

As the feature sizes decrease in deep sub-micron circuit designs, coupling capacitance dominates total capacitance, and crosstalk noise problems become significant and responsible for major timing variations and signal integrity issues.

Timing analysis is an important method to verify that a chip can meet performance requirements. Given a circuit network and its component models, timing analysis calculates signal propagation delay to verify whether the results can be delivered on time at the outputs. Unlike dynamic timing analysis, static timing analysis uses a vectorless approach to analyze the network topology without simulation. Traditional static timing analysis ignores cross coupling effects between wires, or approximates the coupling capacitance by a 2X (Miller factor) grounded decoupled capacitance to account for the worst case delay. This approach not only reduces delay calculation accuracy, but can also be shown to underestimate the delay in certain scenarios. We propose an efficient method to estimate this Miller factor so that the delay response of a decoupled circuit model can emulate the original coupling circuit. Under the assumptions of zero initial voltage, equal charge transfer, and $0.5V_{DD}$ as the switching threshold voltage, an upper bound of 3X for maximum delay and a lower bound of -1X for minimum delay is proven.

Crosstalk coupling is also very sensitive to switching windows, in which signal nets can make transitions. It is the signal switching that causes the wire to inject extra current to its neighboring wires and affect their signal delay or arrival times. Thus, it is important to capture the switching windows for evaluating the crosstalk effect. However, the switching windows again depends on the signal arrival times. The way to resolve this mutual dependency is through iterations. We will build the theoretical foundation to analyze the nature of these iterations considering modeling, accuracy, and mathematical properties and also propose effective ways to converge these iterations. A time slot approach is used to reduce pessimism of crosstalk analysis.

Crosstalk is also subject to functional correlation which is similar to the false path problem (i.e., the neighboring wires might not switch all at the same time in the same direction due to logic correlation). To evaluate a maximum crosstalk noise, we must search and compute the logic condition that produces the maximum peak noise. A conservative approach assumes every net can switch at the same time in the same direction, while the approach we propose can eliminate this false switching combination. A similar idea arises in timing analysis to eliminate false paths. However, the maximum crosstalk problem is even more complicated due to its optimization nature and interaction across a large set of signals.

The goal of this work is to achieve a computationally efficient, accurate, but conservative approach to crosstalk analysis for digital circuits.

Chapter 1

INTRODUCTION

1. Motivation

This work addresses how to analyze the digital noise effects and their impact on static timing analysis (**STA**). These issues have design integrity impact but are typically ignored in older technologies because of the high noise immunity of CMOS circuitry, and the process technologies. However, as the feature size decreases in the deep sub-micron (**DSM**) era, the aspect ratio of metal wires increases (i.e., the thickness of a metal wire is increased and the pitch width is reduced, and as a result, coupling capacitance dominates the total capacitance). Thus, noise is easily coupled from the neighboring nets and becomes a signal integrity issue. Moreover, timing is also affected by this coupling, since it injects extra AC current into a coupled net. These are called **crosstalk effects**.

Traditional chip design focuses on how to meet the timing constraints of a circuit specification and to reduce its area. Thanks to CMOS circuitry, several analog circuit design issues have been avoided or simplified. Therefore, more and more switching components are allowed to integrate onto a single chip, and eventually a system on chip (**SOC**) can be realized. However, the current process technology introduces a large amount of switching digital noise, making the design process more difficult, and more similar to an analog design.

We are motivated to study the effects of digital crosstalk noise, focusing on how to analyze them on the SOC designs, including timing variation and its associated signal integrity issue. The goal of this work is to develop accurate and computationally efficient approaches to calculating the influence of coupling capacitance in static CMOS digital circuits.

1.1   Process Trends

The increasing wire aspect ratio was originally expected to reduce interconnect delay in DSM technology. Figure 1.1 shows the difference of the metal wire profiling between the old technologies and the DSM technologies. The lateral or side-wall capacitance

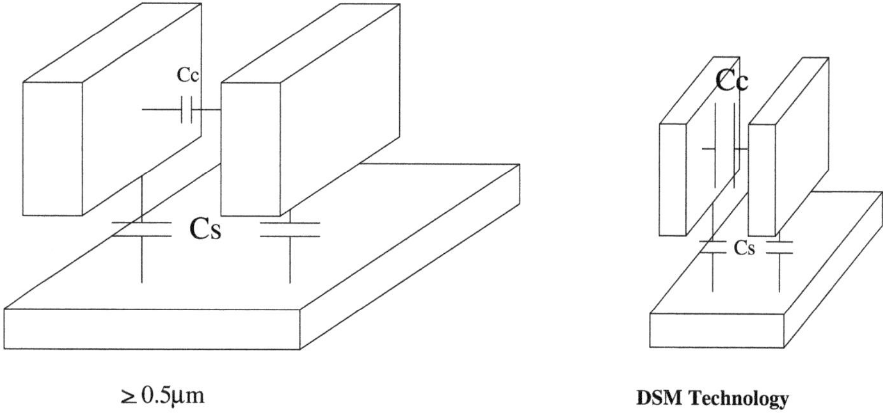

Figure 1.1.   Metal Wire Aspect Ratio Change over Technologies

becomes a dominant factor for the total capacitance calculation. Figure 1.2 (ITRS report, 1998 version) shows the change of the metal aspect ratio over time. As technologies advance, routing pitch decreases and the aspect ratio increases. Note that coupling capacitance over substrate capacitance ratio also increases. This side-wall or coupling capacitance links neighboring nets' switching to a signal net electrically; hence, the crosstalk effects occur naturally.

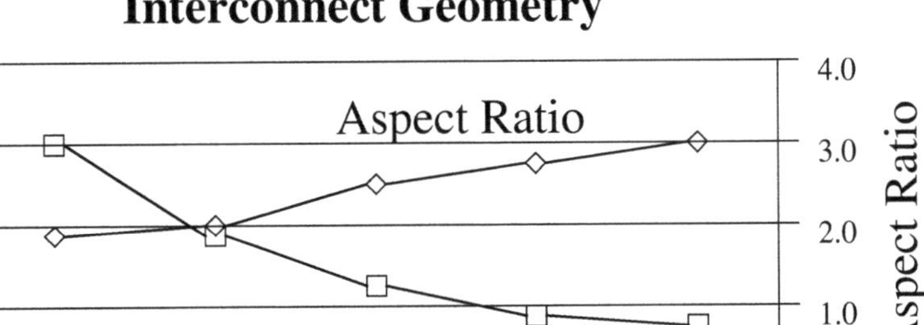

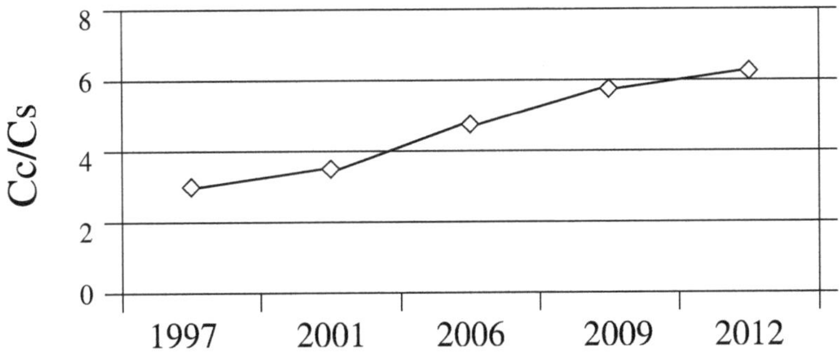

Figure 1.2.   Metal Wire Aspect Ratio Change over Technologies

1.2   CMOS Circuitry

A static CMOS circuit typically has a very good noise rejection capability. Figure 1.3 shows an input-output transfer curve. An input signal is considered logic 1 if the voltage level is above V_{iH}, or logic 0 if the voltage level is below V_{iL}. Due to the CMOS circuitry, input voltage tolerance is rather large. It also implies better noise tolerance compared with analog circuitry. However, if any noise can pull the input voltage up over V_{iL}, the output voltage is easy to saturate and changes into a false state [Kir97].

However, as CMOS technologies lower down the supply voltage as well as the threshold voltage of a transistor, which is equivalent

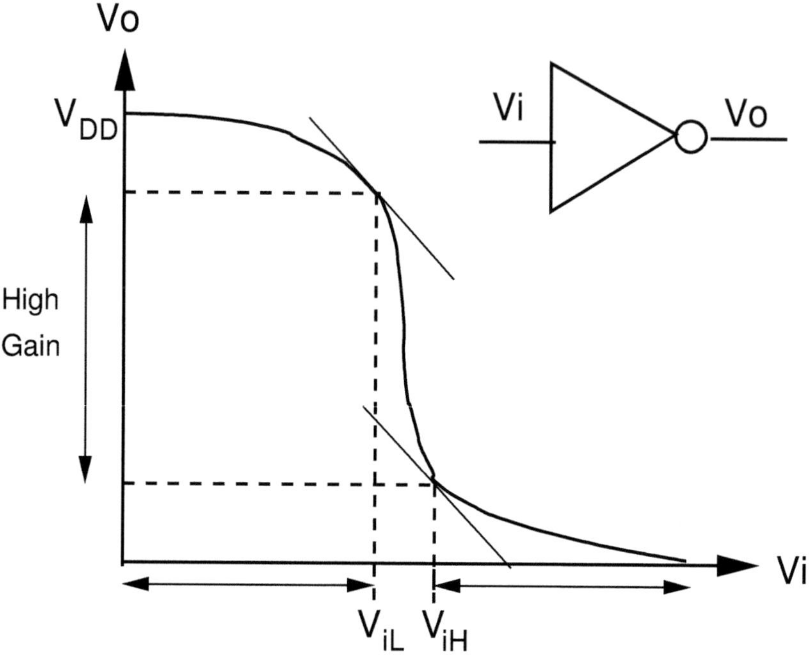

Figure 1.3.   Input-Output Transfer Curve of a CMOS Inverter

to reducing the noise margin, digital designs are more and more susceptible to noise.

Dynamic CMOS circuitry is another circuit family that is sensitive to glitches. Using pre-charging approach, it lacks the capability of static CMOS to restore a logic level back once when there is any coupled noise.

2.   Background and Crosstalk Effects
2.1   Static Timing Analysis

Timing analysis is an important method to verify if a chip can meet performance requirements. Given a circuit network and its component models, timing analysis calculates signal propagation delay to verify whether the results can be delivered on time at the outputs. Unlike dynamic timing analysis, static timing analysis uses a static approach to analyze the network topology without exhaustive simulation of all possible input paths. Although it is conservative and pessimistic, STA is still very efficient even for multi-million-gate designs, compared with dynamic simulation approaches. Therefore, the current design trend employs STA in a standard flow during the chip design process.

2.2   Crosstalk Effects

Conventionally, a **victim** is a net that suffers from noise effects, and an **aggressor** is a net that contributes the noise. These roles can change depending on the context. Consider Figure 1.4, where

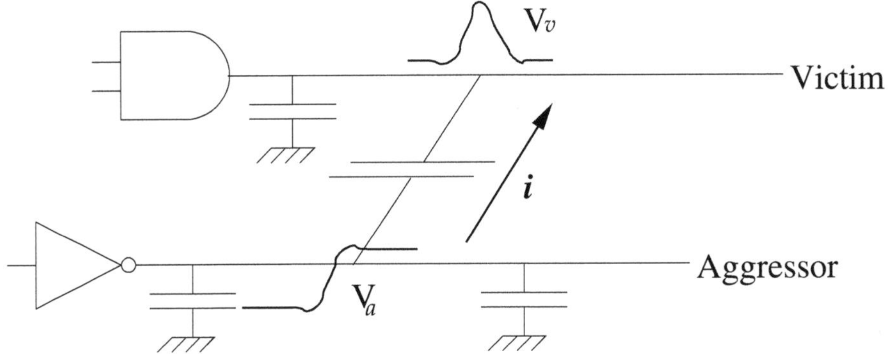

Figure 1.4.   AC current injected from an aggressor

the victim receives AC current from the aggressor, which makes a transition from 0 to 1. The current is written as:

$$i = C \frac{dV_a - dV_v}{dt},$$

where C is the coupling capacitance, and V_a and V_v are the voltage of the aggressor and the victim, respectively.

If the victim is not switching, V_v is restored to its original state, and its waveform becomes a glitch propagating forward. If the victim is switching, depending on the direction of this switching, there is almost no current injecting to the victim (i.e., the coupling C is negligible if dV_v is changing in the same direction as dV_a;

otherwise, the coupling C might have a "double" capacitance effect, if they are switching in the opposite directions). Thus, we have two crosstalk effects: glitch and timing variation. Signal integrity degradation or glitching occurs when the noise couples to a static signal, causing an erroneous logic voltage level. For precharge logic, this is a deleterious effect. Timing variation occurs when the noise couples to a switching signal, causing delay degradation or speed-up. Many approaches have been proposed to take these effects into account in static timing analysis [Sap99, FFP$^+$00, TCE00, ARP00, CKK00b, XCMS00, ZSN01, CKTK02].

2.3   Functional Failure

Crosstalk might induce functional failure or glitch effects[She98a] [Kir97]. For example, considering Figure 1.5, the glitch induced on the victim net could propagate to a latch and change its state, creating a functional failure. Note that glitch propagation is also related a sensitization problem. Typically, a glitch diminishes very quickly across a CMOS gate or it saturates as a full swing waveform quickly depending on the glitch voltage and its waveform width, and the CMOS gate's AC noise rejection characteristics. This kind

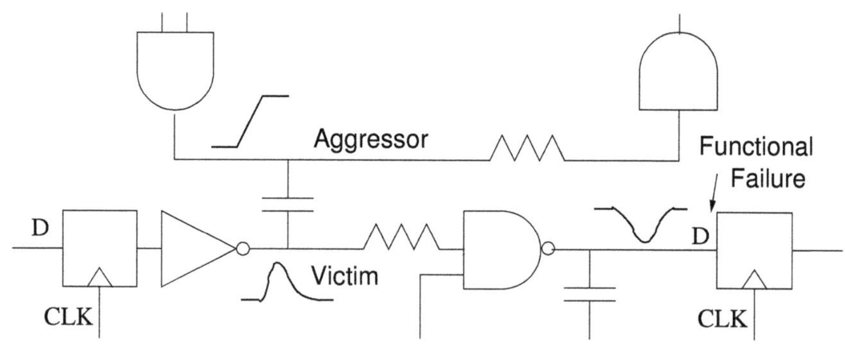

Figure 1.5.   Crosstalk inducing functional failure

of failure is an important issue for signal integrity. Later, we will discuss how to detect the maximum voltage glitch.

2.4 Timing Variation

Crosstalk noise can cause timing variations in two ways. If two nets switch at about the same time in the same direction, the delay is decreased, and for switching in the opposite directions, the delay is increased. Delay decrease is a concern because it may cause hold time violations, while delay increase is also a concern because it my cause setup time violations. Consider Figure 1.6, where the

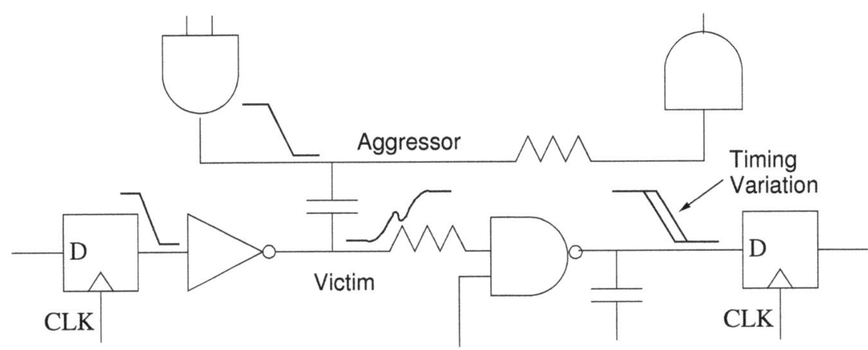

Figure 1.6. Crosstalk inducing timing variation

aggressor's coupling noise can add extra delay to the signal propagation. If the adjacent nets are quiet, there is no delay variation. Therefore, it is important to identify when a net can possibly switch from one state to another. Thus, a **switching window** is defined as a timing duration in which a net or a timing node can possibly make transitions. Figure 1.7 shows an example where node z has several timing paths, one of which creates the earliest arrival time, while the other creates the latest arrival time. The interval between these two time points forms a **switching window**. Identifying overlapping between switching windows can reduce pessimism involved in crosstalk noise analysis, because no timing variation can be induced between two nets when there is no overlap of the switching windows.

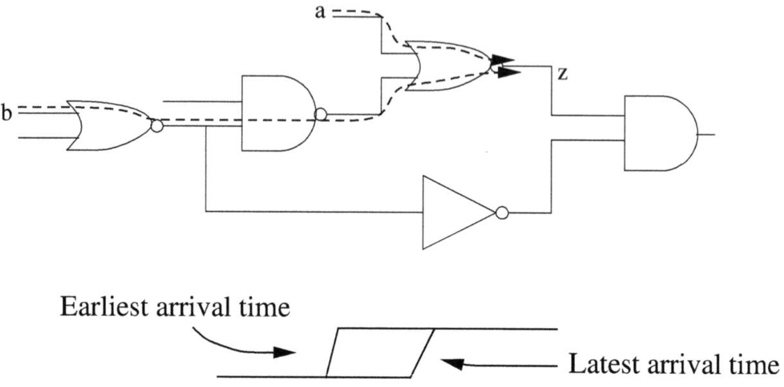

Figure 1.7.   Switching window example

3.   Search Space Pruning

In practice, the space for possible signal interaction of crosstalk coupling noise is too large to search efficiently. The spatial, electrical, temporal, and functional properties of a circuit [Kir97] closely interact to result in significant crosstalk noise. These properties are described in the following sections.

Since the only concern is significant noise which is greater than an allowable margin, these properties can then be used to reduce the search space by excluding some of the conditions which apparently do not cause any significant noise on a victim net.

3.1   Spatial Pruning

Intuitively, two adjacent nets in the layout are potential candidates for capacitive coupling. This gives us a first level of pruning: only adjacent nets may have crosstalk coupling effects. Conventional parasitics extraction provides a very good tool to filter out significant crosstalk effects by extracting coupling capacitance only between any adjacent nets. Any capacitive coupling effect across more than one metal wire is insignificant in modern processes and is typically ignored.

3.2   Electrical Pruning

Using a simplified lumped RC model, the noise level on the victim due to an aggressor can be calculated as Eq. 1.1. Considering a pair of victim and aggressor nets in the physical layout, the waveform on the victim net can be computed as [RIZK94]:

$$V_v(t) = \begin{cases} V_{DD} - V_{DD}\frac{C_x R_v}{t_f}(1 - e^{-t/R_v C_t}), & 0 < t \le t_f \\ V_v(t_f)e^{-(t-t_f)/R_v C_t} \\ \quad + V_{DD}(1 - e^{-(t-t_f)/R_v C_t}), & t > t_f \end{cases} \tag{1.1}$$

where $C_t = C_v + C_x$. See Figure 1.8, where C_v, C_a, C_x are the capacitance of the victim net, aggressor net, and coupling capacitance, respectively; R_v is the resistance of the victim net. If multiple

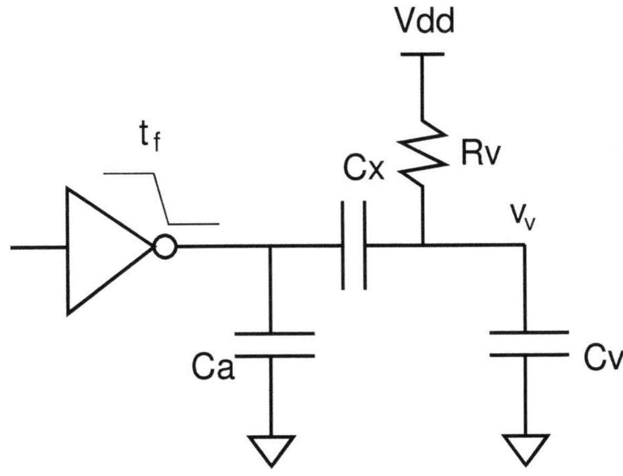

Figure 1.8.   Noise Level Model

aggressor nets have coupling effects on the victim net, C_t should be modified as the sum of C_v and all the coupling capacitances. Assuming linearity, the total coupling effect can be calculated as a superposition of each coupling effect using Eq. 1.1 – that is:

$$\Delta V_v(t) = \sum_i \begin{cases} 0, & 0 < t \leq t_i \\ V_{DD}\dfrac{C_{x_i}R_v}{t_{f_i}}(1 - e^{-(t-t_i)/R_vC_t}), & \\ & t_i < t \leq t_i + t_{f_i} \\ V_{DD}\dfrac{C_{x_i}R_v}{t_{f_i}}(1 - e^{-t_{f_i}/R_vC_t})e^{-(t-t_i-t_{f_i})/R_vC_t}, & \\ & t > t_i + t_{f_i} \end{cases}$$

$$(1.2)$$

where C_{x_i} is each coupling capacitance to the aggressor net i, t_i is the time when the aggressor net i begins to fall, and t_{f_i} is the fall time of the aggressor net i. The maximum value of ΔV_v occurs when all of the $t_i + t_{f_i}$ are equal. This value is used in a *simple worst case* model, where one assumes the extreme case might happen regardless of timing edge alignment or functionality.

Moreover, to address the timing variation due to crosstalk, Chapter 2 provides a method to calculate delay considering coupling based on Miller factors to decouple a coupling circuit model.

3.3   Temporal Pruning

The crosstalk noise is also related to the timing window of each net. For example, in Figure 1.9, the net V's timing window overlaps the net A1's timing window, which means that net V might be attacked by net A1, while the timing window of net A2 is far from that of net V and unlikely to have any crosstalk noise on net V. This approach is reflected in the *static noise analysis* model, where the functional information is ignored but the timing information is included. In this model, if an aggressor signal shares a valid timing window with respect to a victim signal, then one makes the conservative assumption that the transitions are correlated to create maximum crosstalk.

Chapter 3 will provides mathematical formulation and analysis for how to compute these switching windows conservatively, efficiently and robustly considering different coupling models. Chapter 4 also proposes an event-driven algorithm to speed up this com-

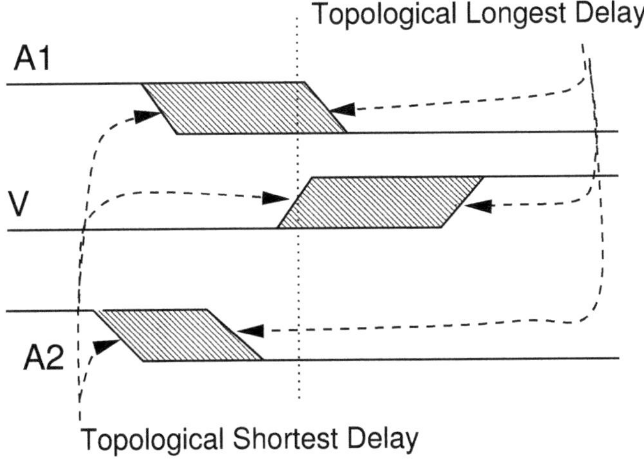

Figure 1.9. Temporal Relationship between Victim Net and Aggressor Nets

putation. Chapter 5 has an approach reducing the pessimism of this static approach.

3.4 Functional Pruning

Applying temporal and electrical pruning alone might be inadequate. Just as functional pruning can be used in static timing analysis to eliminate paths that are never responsible for the delay of the circuit, functional pruning can be used in noise analysis to eliminate those signals/paths that can never be responsible for noise problems because of their functional relationship. By exploring the functional space, we can search for a pair of input vectors whose function allows them to create the maximum noise. This approach is reflected in the *zero-delay model*, where temporal information is ignored but functional information is included. In this model, if an aggressor signal transitions against a victim signal within a clock cycle, then one makes the conservative assumption that the transitions were correlated so as to create maximum crosstalk.

Chapter 6 focuses on how to formulate this problem and translate it into a Boolean constrained optimization problem.

3.5   Problem Complexity v.s. Accuracy

The maximum crosstalk noise depends on the electrical, temporal, and functional aspects of a circuit, so it is generally not easy to be determined exactly. Several techniques have been proposed [SNEZ97b, YCGS97, Kir97] for bounding the maximum crosstalk noise. In general, the electrical information can be easily considered.

In the *floating mode delay* model [DKM93], the prior state of both the victim and aggressor signals are ignored. In this model, both signals are initially assumed to be in an indeterminate state. If there is a vector that causes multiple aggressors to arrive at the same values with temporal correlation, then the conservative assumption is made that the prior state of each signal (before the single vector was applied) will set up the signals for maximum crosstalk. Finally, the fixed-delay 2-vector model considers both timing and functional completely.

The waveform of each method might look like Figure 1.11. In the simple worst case, only electrical pruning is used so that every aggressor is assumed to switch in the opposite direction of the victim net V. The zero-delay model ignores the timing effects, while the static approach ignores the functional effects. The fixed-delay model considers the timing and functional effects at the same time.

The relationship between these techniques is shown in Figure 1.10. The maximum accuracy results from the maximum integration of functional and timing information. Thus the 2-vector approach is more accurate than the others. The relative accuracy between the other approaches varies from circuit to circuit.

Excluding timing information, such as the zero-delay model, we still get a Boolean Constrained Optimization Problem (BCOP), which is equivalent to the Binate Covering Problem (BCP). The static analysis method can result in a linear programming problem [She98a]. However, when the functional aspect is involved, it becomes a BCOP.

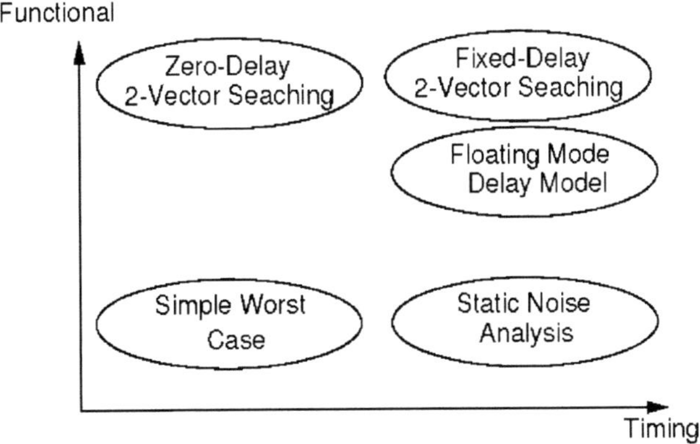

Figure 1.10. Complexity of Crosstalk Noise Analysis

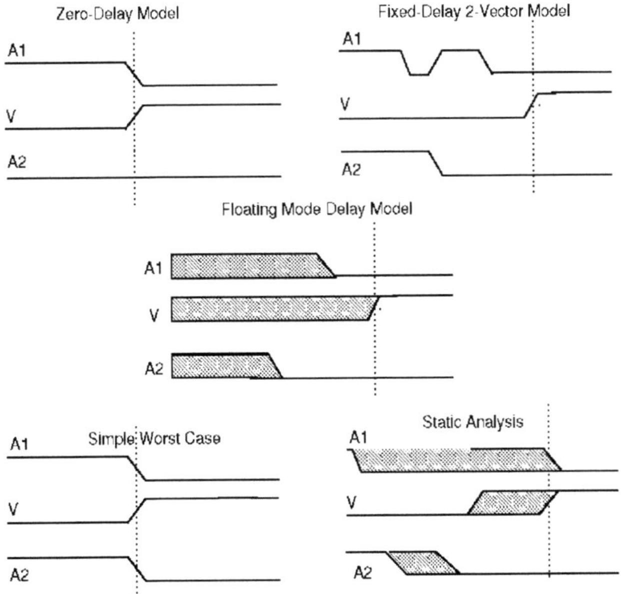

Figure 1.11. Complexity of Crosstalk Noise Analysis

4. Overview

The goal of this work is to develop techniques to analyze digital crosstalk effects with a robust, efficient, accurate and conservative approach for use in STA.

To address delay calculation considering crosstalk effects, in Chapter 2, we propose a method to take this coupling effect into account with a simple extention to traditional delay calculation. A traditional method makes the worst case assumption where twice the coupling capacitance is used to capture this opposite-direction coupling effect. This forms a decoupled version of the circuit for each node, where capacitances are replaced by their Miller equivalent. However, this approach can limit the design space dramatically and lead to a very pessimistic design. Chapter 2 proposes a method based on Miller factors to decouple a circuit such that the traditional approach to calculating timing delay is still valid and offers a good approximation.

In Chapter 3, we focus on the timing aspect, including coupling effects. A formal model is proposed and discussed. The theoretical foundation of switching windows calculation, including coupling noise analysis will be discussed. This is a pure static approach to temporal pruning. Most of the theoretical problems can be solved using this framework.

In Chapter 4, we propose an event-driven algorithm to compute the whole-chip timing in the presence of crosstalk. The event-driven algorithm using different scheduling approaches is compared and explored.

In Chapter 5, we further refine our switching window model into a discrete type to reduce analysis pessimism. Because the number of timing paths is finite, the switching events in a switching window should be discrete. We propose the creation of time slots in a switching window to accurately analyze an aggressor alignment problem.

In Chapter 6, we discuss functional aspects of crosstalk effects analysis. Although the computation comes with high computational complexity, it is still valuable to formulate and solve the functional noise analysis problem. A vector-pair search algorithm is explored to search for the maximum crosstalk noise problem.

Chapter 2

MILLER FACTOR COMPUTATION FOR COUPLING DELAY

In coupling delay computation, a *Miller factor* of more than 2X might be necessary to account for active coupling capacitance when modeling the delay of deep submicron circuitry[YCGS97]. We propose an efficient method to estimate this factor, such that the delay response of a *decoupled circuit model* can emulate the original coupled circuit. Under the assumptions of zero initial voltage, equal charge transfer, and $0.5V_{DD}$ as the switching threshold voltage, an upper bound of 3X for maximum delay and a lower bound of -1X for minimum delay can be proven. Efficient Newton-Raphson iteration is also proposed as a technique for computing the Miller factor or effective capacitance on a coupled model for decoupling. This result is highly applicable to crosstalk coupling delay calculation in deep submicron gate-level static timing analysis. Detailed analysis and approximation are presented. SPICE simulations demonstrate high correlation with these approximations.

1. Introduction

Using a Miller factor is a very convenient method to reduce a highly-coupled circuit to a simpler decoupled approximation (e.g.

a second order system to a first order one). If a coupling capacitor is connected between two nodes, as shown in Figure 2.1, the effective coupling capacitances are equal to $(1-k)C$ and $(1-1/k)C$ grounded capacitance, respectively, where the two nodes have a voltage gain k. In digital designs, $(1-k)$ is conventionally estimated as 2 for the

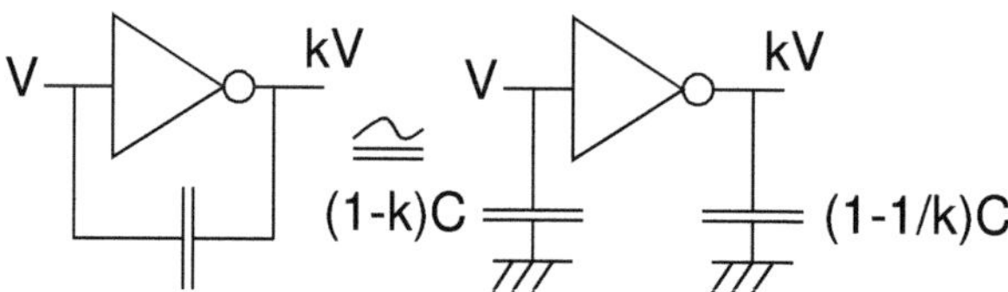

Figure 2.1.   Miller Effect Circuit

opposite direction switching between two coupling nets, and 0 for the same direction switching. However, more than 2X factor has reported [YCGS97]. In this chapter, we present a detailed analysis for how the 2X Miller factor is not an upper bound for coupling delay calculation, and provide a more accurate method to estimate the coupling effect by a decoupling approximation. The factor can be proved as large as 3X under reasonable assumptions. A simple iterative approach and Newton-Raphson iterations are proposed to find these factors. HSPICE simulations are used to demonstrate the accuracy of these factors. Moreover, because the overshoot or undershoot waveform can also affect circuit delay, a correction method is proposed to approximate the effect of overshoot or undershoot.

Our method is crucial for the underlying coupling model for delay calculation in deep submicron circuitry. The Miller factor can be used to approximate a coupling capacitor by two grounded capacitors so that the conventional delay calculation mechanism remains unaltered. Due to the mutual dependency between switching windows and timing delays, the static timing analysis approach must iterate in order to calculate full chip delays in the presence of crosstalk [FFP+00, TCE00]. An efficient and accurate estimation of the coupling effect is crucial for fast convergence. Fixed

Miller factors (2X for the opposite direction switching, and 0 for the same direction switching) are not accurate enough for calculating coupling delay as shown in this chapter. These factors provide neither a bound guarantee nor a good approximation of delay under coupling.

In [YCGS97], the authors show that 2X factor is not an upper bound for crosstalk delay and slew rate, but they do not provide a more accurate factor or prove a new bound. In [DP97], the authors present an iterative algorithm based on [DMP96] to calculate gate delay by approximating the gate response waveform and RC interconnect response. They address how to find an effective capacitance and nonlinear driver model. This can be very accurate for waveform approximation, although it is time-consuming because Newton-Raphson iterations are needed. Moreover, Newton-Raphson iteration with high dimension matrices can be very slow or divergent as it is tricky to find an initial starting point in the convergence region. In contrast with their approach, our approach shows how to decouple the coupling capacitance in a circuit while maintaining delay accuracy. Our method is independent of the driver model. Many analytical models with a linear driver resistance have been proposed, such as [KMV99][BH00][CGB97] [XMS00a]. These models are useful to analyze the crosstalk delay and noise pulse for first level screening. However, as shown in Section 2.2, the linear model might not be accurate enough due to the driver's significant nonlinearity. [SNEZ97a] and [Dev97] report algorithms to calculate the coupling interconnects. In addition, [AKK+00] provides an industrial example of how the crosstalk delay and noise are estimated.

Because the primary use of this work is STA, complete waveform accuracy is not required for static timing analysis – we need only the accuracy up to the switching threshold point to approximate delay accurately. Therefore, we use a decoupling approximation, as shown in Figure 2.1, to emulate and match the circuit response at the switching threshold point. In addition, a decoupling factor like the Miller factor is easy to use and integrate into an existing timing analysis flow.

In this chapter, we introduce a gate driving model and show why superposition or the single driver resistance model is not suitable for crosstalk coupling computation. In Section 2.3, we discuss how to derive a Miller factor for the delay calculation matched at the switching threshold point, and present efficient methods to calculate Miller factors and resolve the convergence issue. Due to the overshoot/undershoot waveform or noise glitch coupled from aggressors, the initial voltage can be quite different from zero. In Section 2.4, we propose a correction factor to fix this problem. It is also useful for glitch waveform estimation. In Section 2.5, we show experimental results from HSPICE simulations.

2.   Gate Driving and Coupling Model

Suppose we have a coupling circuit, as shown in Figure 2.2, where we lump all the interconnect resistance of the victim as R_v,

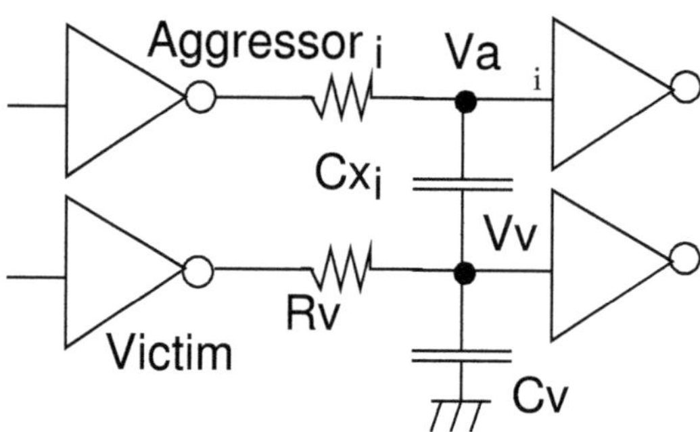

Figure 2.2.   Coupling Circuit

all the grounded interconnect capacitance of the victim net as C_v, and all the coupling capacitance as C_{x_i}. Note that we refer to these two nets as one for victim and the other for aggressor for ease of reference. We calculate the delay impact for a victim net as its aggressor nets contribute noise. Without loss of generality, we assume a rising waveform on the victim net throughout this chapter. It is symmetric for the case with a falling waveform.

2.1 Nonlinearity of Driver Model

Many previous works propose a linear driver model [CGB97] [AKK$^+$00][BH00], in which the driver is connected with a series resistance and a voltage source– for example, a Thevenin equivalent circuit, as shown in Figure 2.3 [DMP96][DP97]. It is useful

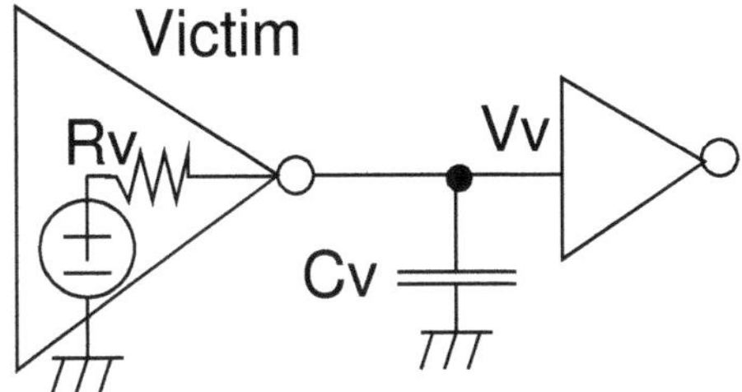

Figure 2.3.   Linear Driver Model

to obtain an analytical formula for delay or noise peak analysis. However, this might not be accurate. In Figure 2.4(a), we show how the driver conductance on a victim net can be nonlinear during the signal transition. It is actually not fixed over time and can vary significantly. The conductance vs. output voltage is also nonlinear, as shown in Figure 2.4(b).

The nonlinear driver resistance prevents superposition of waveforms. In Figure 2.5, we show how a coupling noise can vary with different arrival times of the aggressor. As the aggressor's arrival time is varied, we calculate the coupled noise on the victim net by subtracting the original victim's waveform from the victim's coupled response. If superposition works in these cases, the noise peaks should be the same shape with shifted positions. However, Figure 2.5 shows that the noise peaks can vary according to the arrival times of the aggressor. Therefore, the single resistance assumption for the driver or superposition for waveform estimation might not be accurate for crosstalk delay or noise calculation.

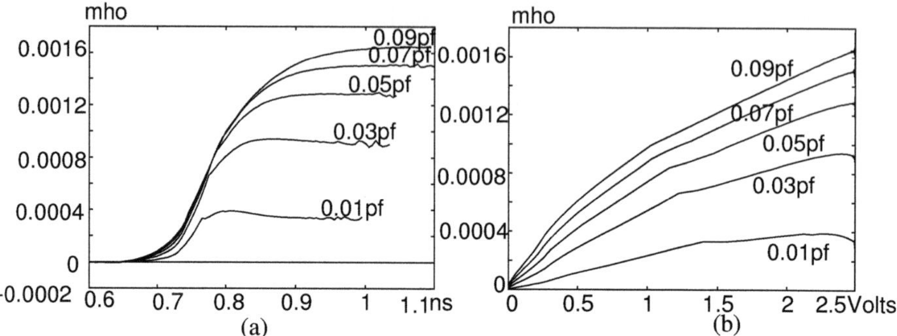

Figure 2.4. (a) Nonlinear conductance varies during signal transition over time for different loadings. (b)Nonlinear conductance vs. output voltage during signal transition for different loadings

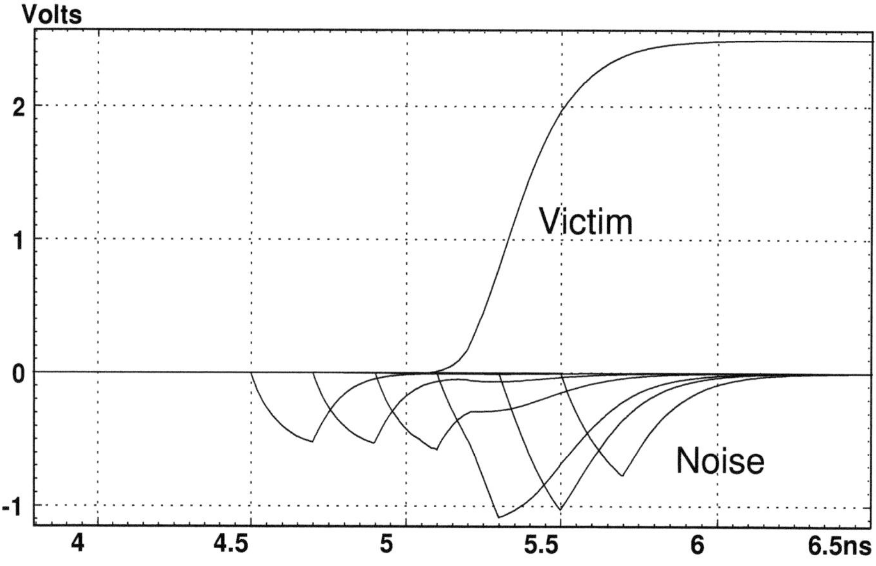

Figure 2.5. Original victim waveform and varying noise due to varying aggressor arrival times

We model the driver resistance as a time-varying and voltage-dependent current source to avoid the inaccuracy of using linear driver model.

2.2 Driver Modeling

We model the time-varying voltage-dependent current source for a victim driver. For a CMOS transistor, the drain current is dependent on the drain-to-source voltage V_{ds} and gate-to-source voltage V_{gs} . It is can be written as $I_{ds} = f(V_{ds}, V_{gs})$. Because the gate-to-source voltage depends on the input waveform, and the drain-to-source voltage is dependent on the output, we model the driving current as a function $I(V, t)$.

3. Decoupling Approximation

It is usually difficult to analyze a coupling circuit like the one in Figure 2.2. If we can replace a coupling circuit with a decoupled circuit using Miller factors to multiply the decoupled capacitors, it is much easier to calculate the delay. Therefore, the objective of decoupling approximation is as follows.

Objective of Decoupling Approximation

The victim's transition interval from 0 to V_{th} is the target transition interval we intend to approximate, and the delay value measured at V_{th} of the transition point (starting from zero voltage) of the decoupled circuit, shown in Figure 2.6, should approximate the response of the original coupling circuit (Figure 2.2). The term $1 - \beta_i$ in Figure 2.6 is called the Miller factor.

In some cases, the initial voltage might not be zero due to the

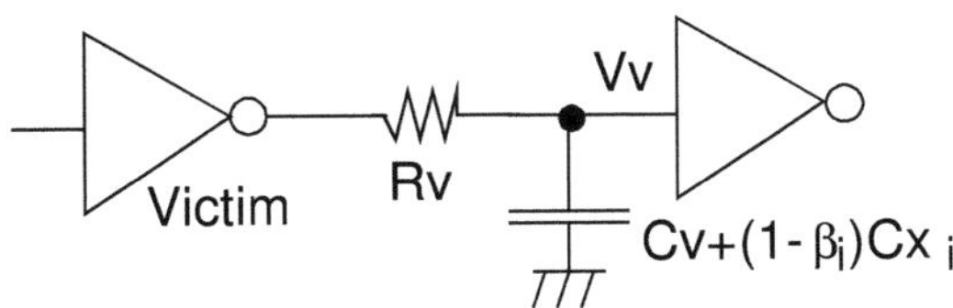

Figure 2.6. Decoupling Approximation Circuit

early arrival of the aggressor signal. We defer this discussion to Section 3.4.

3.1   Coupling Model

Combining the driver model above, the circuit equation for Figure 2.2 can be written as

$$I(V_v, t) = (\sum C_{x_i} + C_v)\frac{dV_v}{dt} - \sum C_{x_i}\frac{dV_{a_i}}{dt}. \qquad (2.1)$$

For simplicity, assume the switching threshold point is at 50% of V_{DD} in this chapter. Integrating it over the period from the rising time of the victim, t_s, to the switching threshold point, t_{th}, for the delay computation of 50% transition, we have:

$$\begin{aligned}
\Delta Q &= \int_{t_s}^{t_{th}} I(V_v(t), t)dt \\
&= \int_0^{V_{th}} (\sum C_{x_i} + C_v)dV_v - \sum \int_{V_{a_i}^0}^{V_{a_i}^1} C_{x_i}dV_{a_i}. \qquad (2.2)
\end{aligned}$$

This can be simplified as, where $\Delta V_v = V_{th}$ and $\Delta V_{a_i} = V_{a_i}^1 - V_{a_i}^0$:

$$\begin{aligned}
\Delta Q &= (\sum C_{x_i} + C_v)\Delta V_v - \sum C_{x_i}\Delta V_{a_i} \\
&= (\sum C_{x_i}(1 - \frac{\Delta V_{a_i}}{\Delta V_v}) + C_v)\Delta V_v \qquad (2.3) \\
&= (\sum C_{x_i}(1 - \frac{\Delta V_{a_i}}{V_{th}}) + C_v)V_{th}. \qquad (2.4)
\end{aligned}$$

Eq. 2.4 implies a factor

$$1 - \frac{\Delta V_{a_i}}{\Delta V_v} = 1 - \frac{\Delta V_{a_i}}{V_{th}}$$

to approximate the coupling capacitance C_{x_i}, assuming *equal charge transfer* at this period. We can find an aggressors' voltage difference in the period of the victim's transition from 0 to V_{th} (e.g. Figure 2.7(a)) to calculate the Miller factor.

3.1.1   Bounds

Assuming equal charge transfer from zero voltage to V_{th} on the victim net, we have a theorem for bounds of the Miller factor for

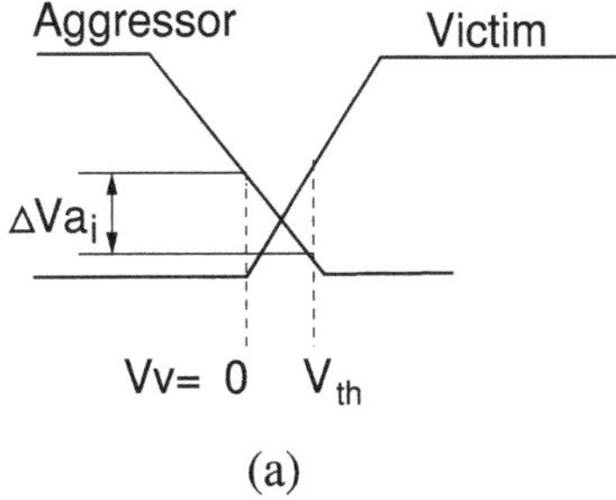

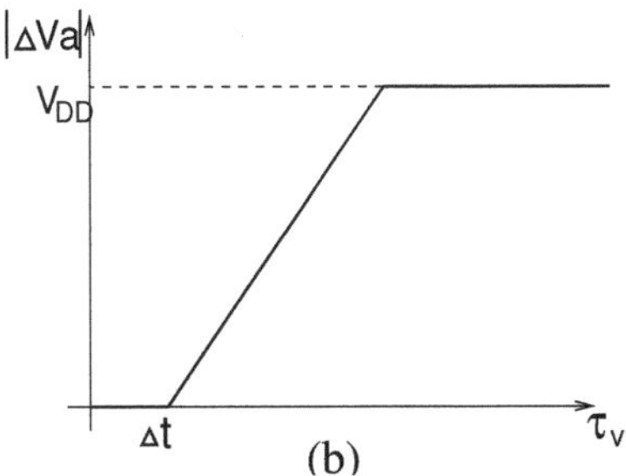

Figure 2.7. (a)ΔV_{a_i} used to calculate the Miller factor for the coupling delay. (b)ΔV_a vs. τ_v, denoted as $G_{\Delta V_a}(\tau_v)$, where Δt accounts for the time difference between the rise(fall) time of the victim and the aggressor.

coupling delay calculation.

Theorem

Under the assumption of zero initial voltage and $V_{th} = 0.5V_{DD}$, the maximum Miller factor of the opposite direction switching is 3 for coupling delay calculation measured at the 50% transition. The minimum Miller factor is -1 for the same direction switching at the 50% transition.

Proof: From Eq. 2.4, the bounds are easily derived. If the aggressor and the victim switch in the opposite directions, we can have an upper bound 3, when ΔV_{a_i} is equal to $-V_{DD}$ and $\Delta V_v = 0.5V_{DD}$.

If the aggressor and the victim switch in the same direction, we can have as a lower bound of -1, when ΔV_{a_i} is equal to V_{DD} and $\Delta V_v = 0.5 V_{DD}$. $\qquad\qquad\square$

3.2 Simple Iterative Approach

ΔV_{a_i} is actually unknown before the coupling delay is calculated, so the Miller factor $1 - \frac{\Delta V_{a_i}}{\Delta V_v}$ cannot be estimated accurately before the delay is calculated. A simple approach can iterate on ΔV_{a_i} until it converges. For ease of notation, we drop the index i for the following discussion wherever there is no confusion. Consider a fixed ramp $V_a(t)$ waveform on an aggressor and the victim switches at the opposite direction. ΔV_a versus the victim's ramp time τ_v is shown in Figure 2.7(b). Some notations are defined as:

- $\beta = \frac{\Delta V_a}{\Delta V_v} = \frac{\Delta V_a}{V_{th}}$,

- the effective capacitance C_{eff}, which is

$$C_{eff} = (1 - \beta)C_x + C_v, \qquad (2.5)$$

- $\tau_v(C_{eff})$ as the τ_v function of effective capacitance, which is the ramp time or slew rate response at the node that the coupling delay measured at the switching threshold point is to be matched, and

- $G_{\Delta V_a}(\tau_v)$ as the ΔV_a function of τ_v, as shown in Figure 2.7(b).

So a composite function $H_\beta(C_{eff})$ can be defined to compute a new β given C_{eff}

$$\begin{aligned}
\beta = H_\beta(C_{eff}) &= \frac{\Delta V_a}{\Delta V_v} = \frac{\Delta V_a}{V_{th}} = \frac{1}{V_{th}} G_{\Delta V_a}(\tau_v) \\
&= \frac{1}{V_{th}} G_{\Delta V_a}(\tau_v(C_{eff})). \qquad (2.6)
\end{aligned}$$

Simple iterative approach just combines Eq. 2.5 and Eq. 2.6, i.e.

$$C^*_{eff} = (1 - H_\beta(C_{eff}))C_x + C_v \qquad (2.7)$$

to iterate until it converges. We note that the convergence rate of this approach is linear.

3.2.1   Convergence of the Simple Iterative Approach

The curves for the Miller factor versus the effective capacitance C_{eff} shows how this algorithm converges. Four possible curves are shown in Figure 2.8. The dashed line is the mapping from the Miller factor calculated by Eq. 2.6, and the solid line is the Miller factor by Eq. 2.5. Starting from the Miller factor equal to 3, Figure 2.8(a),

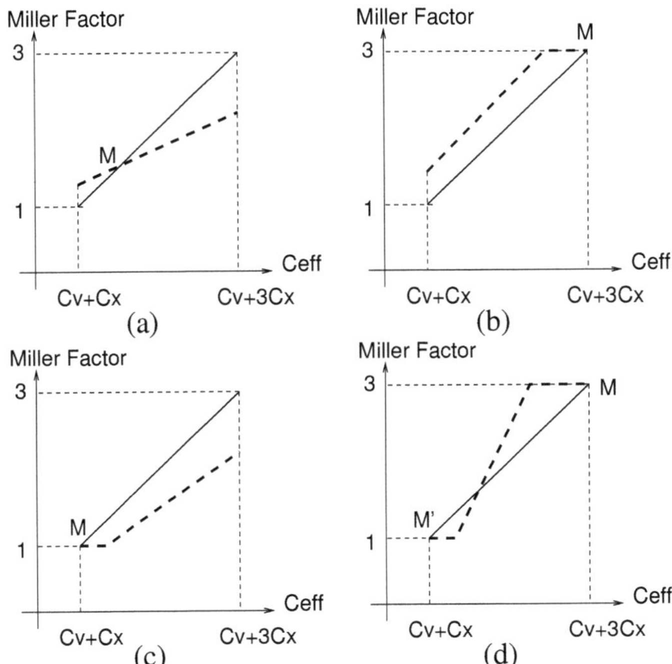

Figure 2.8.   Miller factor vs. effective capacitance, assuming $\Delta V_v = V_{th} = 0.5 V_{DD}$

(b) and (c) can converge to point M. The algorithm is equivalent to starting from an initial Miller factor, mapping it to an effective capacitance along the solid line (Eq. 2.5), and then computing a new Miller factor according to Eq. 2.6 along the dashed line. This process repeats until it converges to point M. In (d) there might be 2 points M and M' that the algorithm can converge to, depending on the initial point. The conservative approach is to take the initial value as 3 to converge to the upper point M. Physically, this condition corresponds to a very weak victim driver and a very sharp aggressor transition. Most of the cases result in a spike, which is not easy to approximate by this method.

3.3   Newton-Raphson Iteration for Miller Factor

The convergence rate can be improved by using the Newton-Raphson iteration. The key is to compute the derivative of β. By using Eq. 2.7, the Newton-Raphson iteration is set to find the root of:

$$f(C_{eff}) = (1 - H_\beta(C_{eff}))C_x + C_v - C_{eff} = 0.$$

The derivative of $H_\beta(C_{eff})$ is:

$$H'_\beta(C_{eff}) \;=\; \frac{1}{V_{th}}G'_{\Delta V_a}(\tau_v)\tau'_v(C_{eff}) \tag{2.8}$$

$$=\; \frac{1}{V_{th}}\frac{d\Delta V_a}{d\tau_v}\frac{d\tau_v}{dC_{eff}}, \tag{2.9}$$

where $\tau'_v(C_{eff})$ can be calculated and found by a simple table lookup from the conventional characterization data of a standard cell library. The Newton-Raphson iteration can therefore be written as:

$$C^*_{eff} = C_{eff} - \frac{(1 - H_\beta(C_{eff}))C_x + C_v - C_{eff}}{-C_x H'_\beta(C_{eff}) - 1}. \tag{2.10}$$

Note that if $H'_\beta(\beta)$ is equal to 0, simplifying Eq. 2.10, we have:

$$C^*_{eff} = (1 - H_\beta(C_{eff}))C_x + C_v.$$

This is equivalent to the simple iterative approach from Eq. 2.7.

The strength of this approach is the quadratic convergence rate of Newton-Raphson iteration and the convergent initial value is easy to find because the value of β is between -2 to 2 if $V_{th} = 0.5V_{DD}$, and hence the value of C_{eff} is between $-C_x + C_v$ and $3C_x + C_v$.

Note that strictly speaking, $G_{\Delta V_a}$ (defined in Section 2.3.2) does not only depend on τ_v, but also depends on the relative delay between the aggressor and the victim, which is in turn dependent on C_{eff}. There is no explicit analytic formula available to accurately describe the relation. Conventionally, $\tau_v(C_{eff})$ is described by a two dimensional table pre-characterized for the driving cells. However, the derivative can be relaxed by finite difference approximation. Empirically, the derivative function $H'_\beta(C_{eff})$ can be well approximated by the finite difference because the original function $H_\beta(C_{eff})$ is very close to a linear function.

The number of iterations versus the ratio of coupling capacitance to grounded capacitance for two approaches are compared and shown in Figure 2.9, where the convergence criterion is such

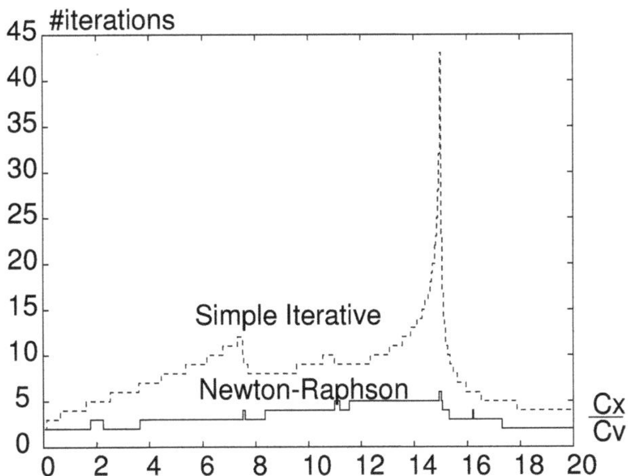

Figure 2.9. Comparison of the number of iterations of Simple Iterative Approach and Newton-Raphson iteration

that the relative error is less than 10^{-4}. The peak in Figure 2.9 is a case for which the simple iterative approach takes 43 iterations to converge. The trace for both approaches for this special case is shown in Figure 2.10. It shows that the simple iterative approach has to iterate more steps around the convergent point, while Newton-Raphson takes few steps to converge.

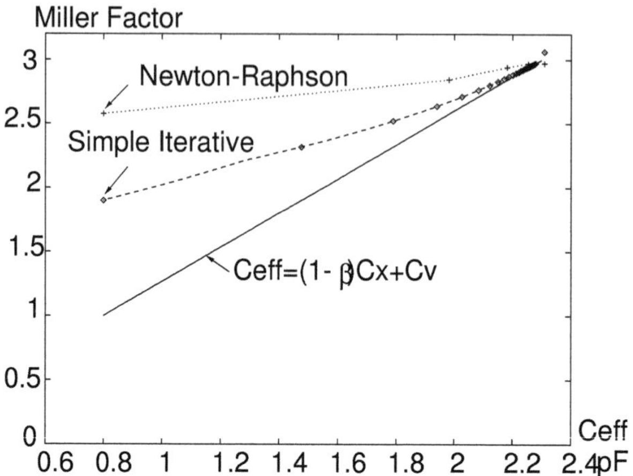

Figure 2.10. A case for Simple Iterative Approach to take 43 iterations to converge.

3.4 Multiple Miller Factors for Multiple Coupling Nets

For multiple nets coupling together, the simple iterative approach becomes

$$C_{eff}^{i*} = \sum_k (1 - H_{\beta_k}^i(\mathbf{C}_{eff}))C_{x_k}^i + C_v^i. \qquad (2.11)$$

However, multiple C_{eff} values can be found in a single Newton-Raphson run. There is one C_{eff} for each coupling net. Once C_{eff} is known, ΔV_{a_i} can be computed, and hence the Miller factors. Unlike the approach in [DP97], we address how to use a decoupled circuit to emulate the original one, while [DP97] addresses how to find C_{eff} and the gate delay. Our model is independent of the driver model used. Actually, the gate driver model and RC interconnect delay are encapsulated in the function $\tau_v(C_{eff})$, where C_{eff} and τ_v are the effective capacitance and ramp time, respectively, at the node where the coupling delay should be matched.

For completeness of this chapter, we derive the Newton-Raphson iteration in the following equations. Suppose we have:

$$f^i(\mathbf{C}_{eff}) = \sum_k (1 - H_{\beta_k}^i(\mathbf{C}_{eff}))C_{x_k}^i + C_v^i - C_{eff}^i = 0$$

set to find the roots, where the superscript i denotes the related variable at net i, and net i has net x_k as a coupling net with coupling capacitance $C_{x_k}^i$. The partial derivative of $f^i(\mathbf{C}_{eff})$ is

$$\frac{\partial f^i(\mathbf{C}_{eff})}{\partial C_{eff}^j} = \begin{cases} -1 - \sum_k C_{x_k}^i \dfrac{\partial H_{\beta_k}^i(\mathbf{C}_{eff})}{\partial C_{eff}^i} & \text{if } i = j \\ -\sum_k C_{x_k}^i \dfrac{\partial H_{\beta_k}^i(\mathbf{C}_{eff})}{\partial C_{eff}^j} = -C_{x_j}^i \dfrac{\partial H_{\beta_j}^i(\mathbf{C}_{eff})}{\partial C_{eff}^j} & \text{if } i \neq j \end{cases},$$

where

$$\frac{\partial H_{\beta_k}^i(\mathbf{C}_{eff})}{\partial C_{eff}^j} = \frac{1}{V_{th}} \frac{\partial \Delta V_{a_k}^i}{\partial C_{eff}^j} = \frac{1}{V_{th}} \frac{\partial G_{\Delta V_{a_k}^i}(\tau_v^i, \tau_v^k)}{\partial C_{eff}^j},$$

which is nonzero only when $i = j$ or $j = k$.

We test this algorithm on a case of two coupling nets with a rising ramp on both nets. The result is shown in Figure 2.11, where C_v is equal to 0.05pF. Note the difference between this and the previous section is that neither of the nets has a fixed output waveform. The simple iterative approach fails to converge within 200 steps below the capacitance ratio 1.75. We also test a case with four coupling

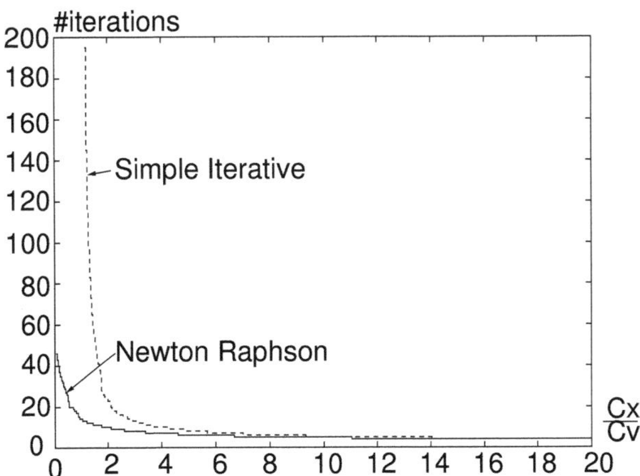

Figure 2.11. Comparison of the number of iterations for Simple Iterative Approach and Newton-Raphson Iterations with 2 coupling nets.

nets with a rising ramp on each net. The result is as shown in Figure 2.12.

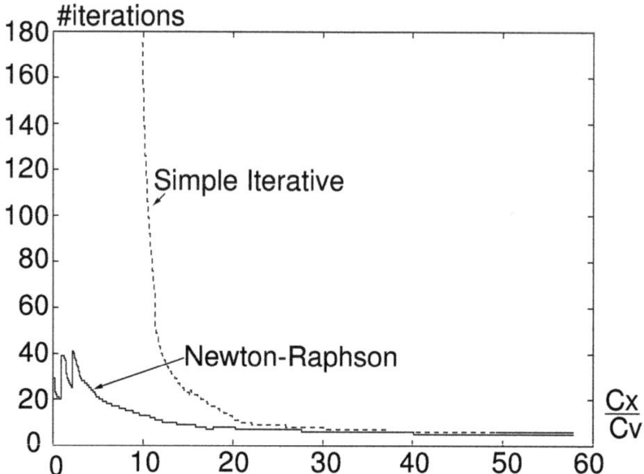

Figure 2.12. Comparison of the number of iterations for Simple Iterative Approach and Newton-Raphson Iterations with 4 coupling nets.

3.5 Slew Rate (Transition Time) Calculation

Slew rate (transition time or ramp time) is another factor that affects the delay calculation. However, if the effective coupling factors are used to approximate the slew rate, we have to match up to the upper point of transition (typically 90% or 80% of transition) or down to the lower point of transition (typically 10% or 20% of transition). It might need more than one Miller factor to calculate. The difference is to make $\Delta V_v = 0.9V_{DD}$ in Eq. 2.4 for the upper point of transition at 90% of V_{DD}, or $\Delta V_v = 0.1V_{DD}$ for the lower point of transition at 10% of V_{DD}.

4. Nonzero Initial Voltage Correction

Some waveforms might not start from zero voltage. This leads to another source of errors for the Miller factor approximation. We show how to correct this problem in this section.

4.1 Glitch Waveform Approximation

Consider a case when an aggressor is making transition before a victim, such that an undershoot waveform occurs on the victim net, as shown in Figure 2.13(a). The victim's initial voltage being zero

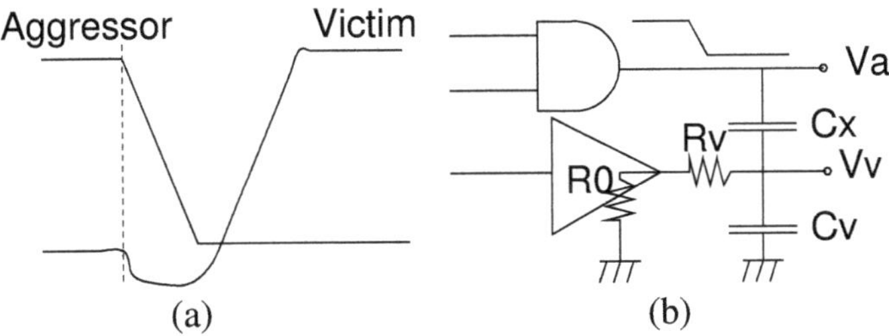

Figure 2.13.   (a)Undershoot Waveform (b)Undershoot Circuit

is not exactly accurate due to the glitch coupled from the aggressor. Consider a falling aggressor with a ramp waveform. A simplified model is used as shown in Figure 2.13(b). We have

$$\frac{V_v}{R_0 + R_v} + C_x \frac{d}{dt}(V_v - V_a) + C_v \frac{dV_v}{dt} = 0.$$

Assume $V_a = (1 - t/\tau_a)V_{DD}$ and $\frac{dV_v}{dt} = \frac{V_v}{t}$ by averaging approximation. It can be rewritten as

$$\frac{V_v}{R_0 + R_v} + C_x\left(\frac{V_v}{t} + \frac{V_{DD}}{\tau_a}\right) + C_v \frac{V_v}{t} = 0$$

$$V_v = \frac{-\frac{C_x}{\tau_a}V_{DD}t}{t/(R_0 + R_v) + C_x + C_v}, \qquad t = 0 .. \tau_a. \qquad (2.12)$$

If multiple aggressors are present, and the initial voltage of V_v is nonzero V_v^0, we can extend it as

$$V_v = \frac{-t \sum \frac{\delta_i C_{x_i}}{\tau_{a_i}} V_{DD} + (\sum C_{x_i} + C_v)V_v^0}{t/(R_0 + R_v) + \sum C_{x_i} + C_v} \qquad (2.13)$$

where δ_i is 1 if the aggressor is falling, or -1 if the aggressor is rising, and zero otherwise. This equation also applies when multiple aggressors join in different time points.

This modeling is verified with HSPICE simulations. One example is shown in Figure 2.14. The victim's driver resistance is 856 ohm, the aggressor has a perfect falling slope of 0.5ns, Cx is 0.02pf, Cv is 0.01pf, and V_{DD} is 3.3v. It shows that at the corners of the curve this model has some small errors.

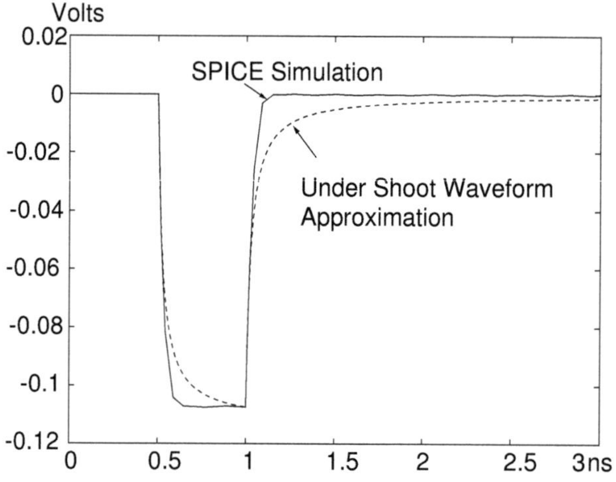

Figure 2.14.   Experiment for Undershoot Modeling

5.   Experimental Results

We verify the estimated Miller factors by HSPICE simulation on a simple circuit with a fixed ramp input on an aggressor net, and a pure capacitive loading on the victim net with some coupling capacitance. In Figure 2.15, we vary the aggressor's arrival time to see the effect of the delay variation on the victim net. The estimated Miller factor is calculated using the decoupling approximation, described in Section 2.3, and the undershoot correction is also computed. Using this factor, HSPICE simulation is performed on the decoupled circuit again to measure the delay. The same procedure is repeated for 2X Miller factor. Our method (marked as decoupling approximation) closely follows the original coupling circuit within 7.5%, while the 2X Miller factor can be farther(18.2%) off. In addition, it shows that 2X is not an upper bound. In Figure 2.16, we vary

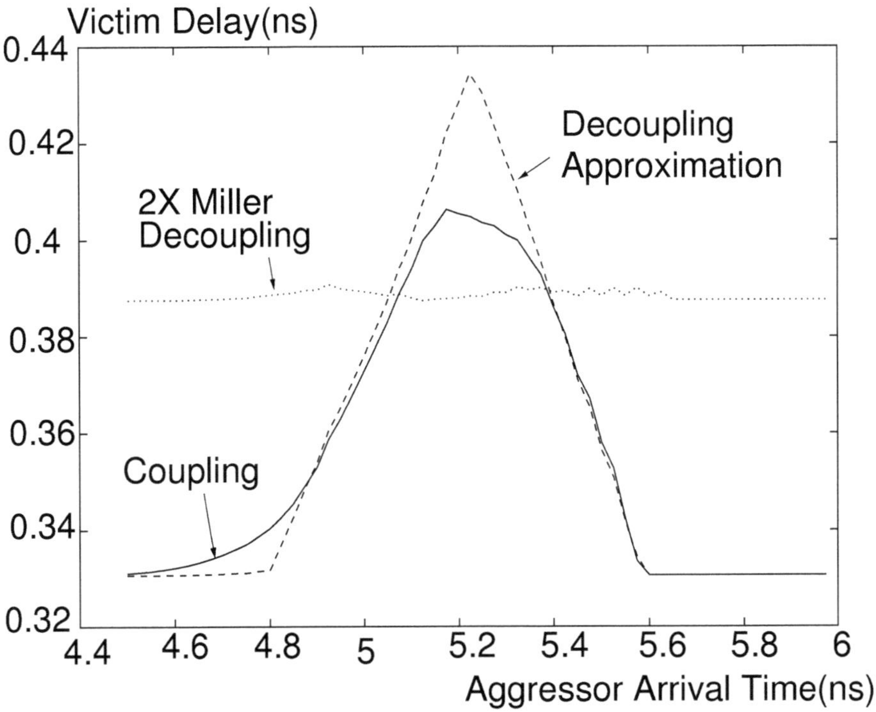

Figure 2.15.   Victim delay vs. aggressor's arrival time

the aggressor's ramp time to see the effect of delay variation on
the victim net. Our method (marked as decoupling approximation)
closely follows the original coupling circuit, while the 2X Miller
factor can be farther off. Typical waveform response is shown in
Figure 2.17.

6.   Review of Conservativism

The approach proposed in this chapter is a good approximation to
coupling delay, although it is not strictly conservative (see Figure
2.15). Some of the approximated delays may under-estimate or
over-estimate. It may be combined with temporal pruning methods
as described in Chapter 3 and 4 to give better accuracy.

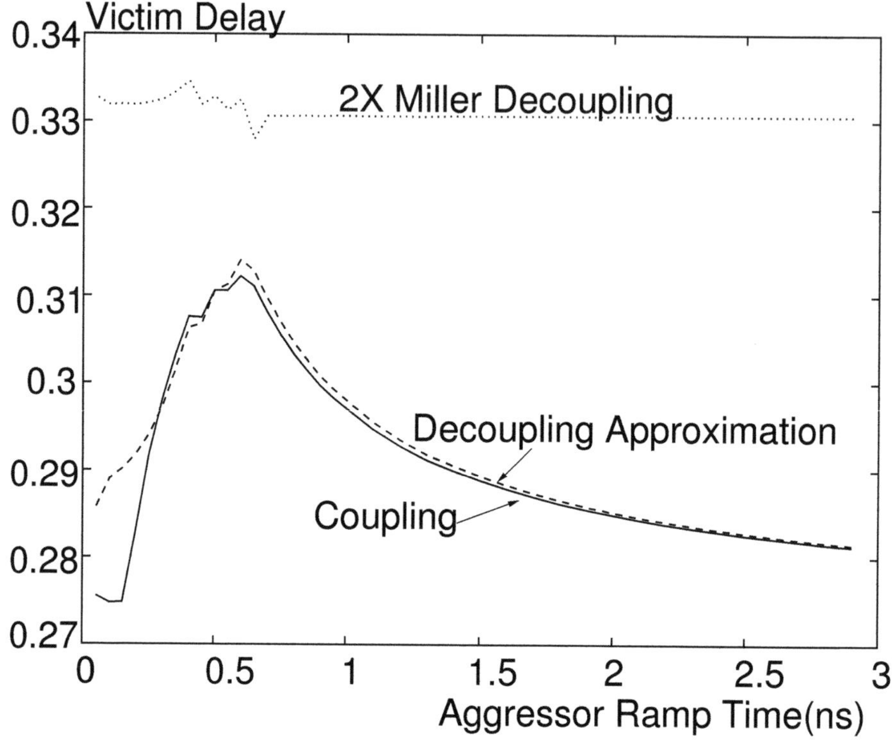

Figure 2.16.   Victim delay vs. aggressor's ramp time

7.   Conclusion

In this chapter, we have proposed a simple and accurate method to estimate the Miller factor for approximating a coupling circuit by a decoupled circuit. It is well-suited for coupling delay calculation in very deep submicron designs. An efficient Newton-Raphson method is proposed to find the Miller factors or effective capacitance. In addition, we prove an upper bound of 3X for the opposite direction switching, and a lower bound of -1X for the same direction switching. The conventional 2X factor is shown clearly not to be a bound and can be very inaccurate for coupling delay calculation.

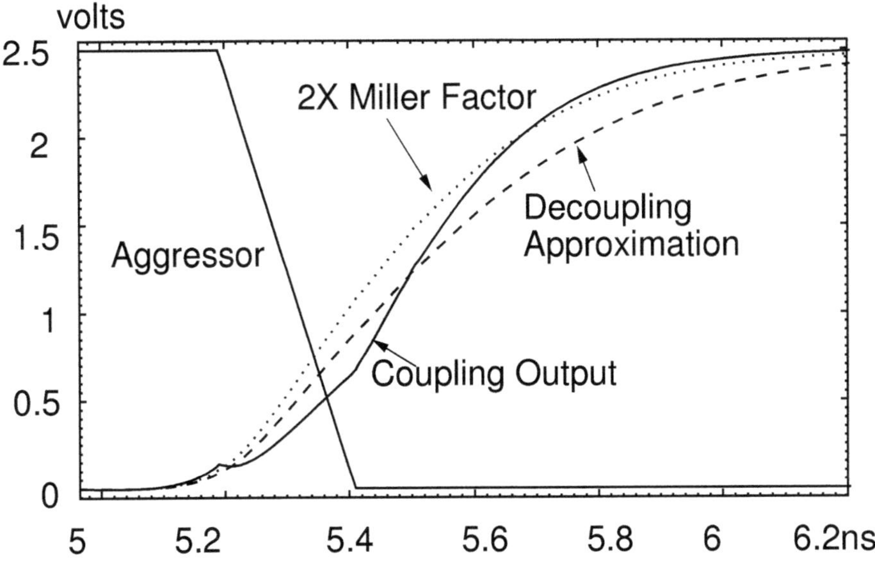

Figure 2.17. Example of waveform response

Chapter 3

CONVERGENCE OF SWITCHING WINDOW COMPUTATION

Detecting the overlap of switching windows between coupled nets is an important static technique to accurately locate crosstalk noise. The amount of coupling noise depends on extent of overlap between switching windows, but the coupling noise also affects signal switching times (and therefore switching windows). Hence, there is a mutual dependency between switching windows, so computing the coupling effect requires iterations to converge. In this chapter, we discuss the issues and provide answers to the important questions involved in convergence and numerical properties, including the effect of coupling models, multiple convergence points, convergence rate, computational complexity, non-monotonicity, continuity and the effectiveness of bounds. Numerical fixed point computation results are used to explain these properties. Our contribution here builds a theoretical foundation for static crosstalk noise analysis.

1. Introduction

With crosstalk noise, switching windows are considered mutually dependent in static timing analysis (STA), and the computation cannot be completed in a single traversal of nets in general

[Sap99, FFP$^+$00, TCE00, ARP00, CKK00b, XCMS00, ZSN01]. Iterations are therefore needed to resolve the mutual dependency. Thus, the following questions arise:

- Does it always converge? What coupling or overlapping models lead to divergence?

- Is there a unique convergence point independent of an initial condition? Is it physically realizable or just a bound?

- At most how many iterations are needed? What is the computational complexity? How fast does the process converge?

- If a gate delay is reduced, is a circuit's longest path delay considering crosstalk noise reduced as well?

- If a gate delay is increased continuously, will the crosstalk noise also be monotonically increased?

In [ZSN01], the authors suggest the use of a lattice theory to prove convergence of switching windows computation and show that there are multiple convergence points depending on the initial condition. However, the coupling model they used is very primitive and is not accurate due to its discrete nature. We will tackle this problem from a different point of view using a *numerical* fixed point computation perspective [EMU96]. We will also examine the impact of discrete and continuous coupling models on convergence and numerical properties of switching windows computation. Moreover, the questions listed above will be studied in this chapter.

The remainder of this chapter is organized as follows. In Section 3.2, we introduce some definitions used in the chapter, and give the upper and the lower bounds. In Section 3.3, we address fixed point computation applied to switching windows computation. In Section 3.4, we examine the underlying models used in switching windows computation. In Section 3.5, we discuss convergence and efficiency issues.

2. Background

A quantity is said to be **noisy** if crosstalk noise effects have been included. In contrast, a quantity is said to be **noiseless** or **nominal** if the crosstalk noise effects are not included. For example, a **noisy delay** is the nominal delay plus the extra delay induced by crosstalk effects. Similarly, a **noisy switching window** is the nominal switching window including the extra path delay (timing variation) induced by crosstalk effects.

A **coupling edge** exists from the victim to the aggressor in the STA timing graph if there is a coupling capacitance linked between them. A coupling edge is said to be **active** if the delta delay induced by this edge has been included in its victim's noisy switching window. In the process of switching windows computation, a coupling edge can change its **state** from inactive to active due to overlapping of noisy switching windows, or vice versa.

For simplicity, we assume a single delay value on interconnect for all of the fanouts of each net in the following discussion. Let x_i and y_j be the latest arrival time of net i and the earliest arrival time of net j, respectively. Let $\theta_{ij} : R \to [0,1]$ be a switching window

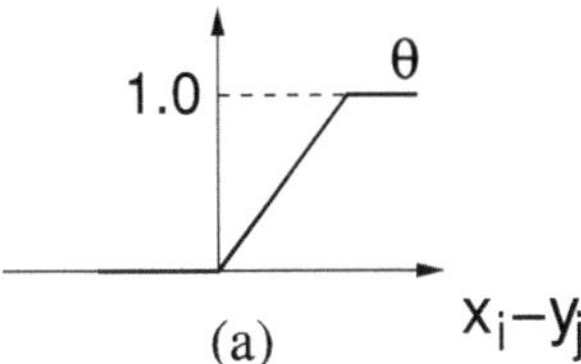

Figure 3.1. θ function to represent the overlapping

overlapping function of time difference. Figure 3.1 is an example of function $\theta_{ij}(x_i - y_j)$. If $x_i - y_j < 0$, the switching windows do not overlap, and resulting in no coupling noise. As $x_i - y_j > 0$, the coupling noise begins to show and eventually saturates at a normalized value of 1.0. The delta delay induced from aggressor net j to victim net i is thus written as:

$$\theta_{ij}(x_i - y_j)\Delta t_{ij}^{max},$$

where $\Delta t_{ij}^{max} > 0$ is the maximum delta delay induced from aggressor net j to victim net i.

An index set denotes a collection of net indices, representing a subset of all nets. Let d_i be the interconnect delay of net i, A_i be an index set of aggressor nets of net i, D_i be an index set of fanin nets of net i's driving gate, and δ_{ik}, $k \in D_i$, be the gate delay from a fanin net k to net i. Figure 3.2 relates these variables to a circuit

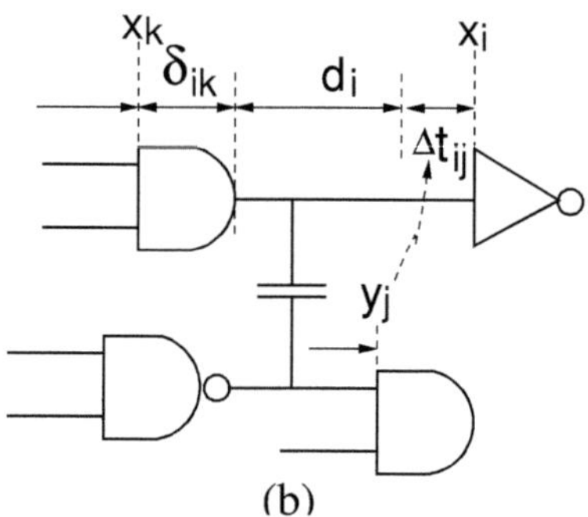

Figure 3.2.   Variables to represent delays in a coupling subcircuit

diagram. x_i and y_i are written as

$$x_i = \sum_{j \in A_i} \theta_{ij}^{late}(x_i - y_j)\Delta t_{ij}^{max} + d_i + \max_{k \in D_i}(x_k + \delta_{ik}) \qquad (3.1)$$

and

$$y_i = -\sum_{j \in A_i} \theta_{ij}^{early}(x_j - y_i)\Delta t_{ij}^{max} + d_i + \min_{k \in D_i}(y_k + \delta_{ik}), \qquad (3.2)$$

respectively. Note that θ_{ij}^{late} and θ_{ij}^{early} are not exactly the same function. To simplify the notation, we denote it as θ_{ij}, because it can be distinguished by the context to use θ_{ij}^{late} for x_i or θ_{ij}^{early} for y_i, respectively.

A quantity associated with no coupling effect is unconditionally realizable. For example, a noisy arrival time in Eq. (3.1) of a victim net is said to be **realizable** if y_j's and x_k's are all realizable. Starting from inputs of a circuit, the longest and the shortest noisy delay to each net have to be realizable in order to achieve realizable switching windows for the entire circuit.

2.1 Simple Upper and Lower Bounds for Switching Windows

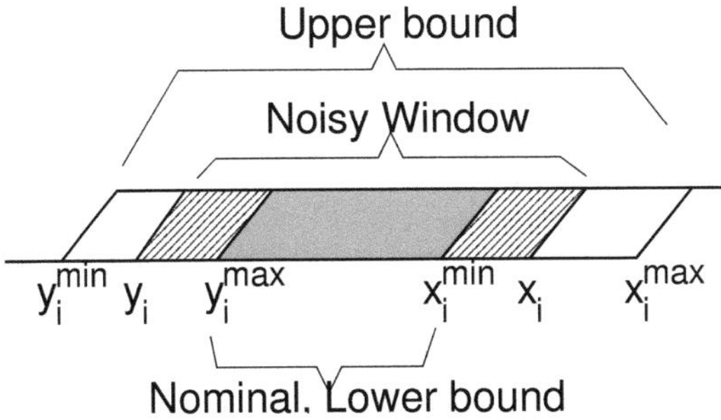

Figure 3.3. Bounds for switching window

What are the possible latest and earliest arrival times that consider crosstalk noise? When all crosstalk noises are active and induce the maximum extra delay to increase the latest arrival time and reduce the earliest arrival time, this is the largest switching window (i.e. **the upper bound**), that can be achieved with crosstalk noise. The upper bounds are written as

$$x_i^{\max} = \sum_{j \in A_i} \Delta t_{ij}^{\max} + d_i + \max_{k \in D_i}(x_k^{\max} + \delta_{ik}) \qquad (3.3)$$

$$\geq \sum_{j \in A_i} \theta_{ij}(x_i - y_j)\Delta t_{ij}^{max} + d_i + \max_{k \in D_i}(x_k + \delta_{ik}), \qquad (3.4)$$

and

$$y_i^{\min} = -\sum_{j \in A_i} \Delta t_{ij}^{\max} + d_i + \min_{k \in D_i}(y_k^{\min} + \delta_{ik}) \qquad (3.5)$$

$$\leq -\sum_{j \in A_i} \theta_{ij}(x_j - y_i)\Delta t_{ij}^{max} + d_i + \min_{k \in D_i}(y_k + \delta_{ik}). \qquad (3.6)$$

On the contrary, when all crosstalk noises are excluded, the nominal switching window gives **the lower bound** (i.e. the smallest possible switching window). The lower bounds are written as

$$x_i^{\min} = d_i + \max_{k \in D_i}(x_k^{\min} + \delta_{ik}) \qquad (3.7)$$

$$\leq \sum_{j \in A_i} \theta_{ij}(x_i - y_j)\Delta t_{ij}^{max} + d_i + \max_{k \in D_i}(x_k + \delta_{ik}), \qquad (3.8)$$

and

$$y_i^{\max} = d_i + \min_{k \in D_i}(y_k^{\max} + \delta_{ik}) \tag{3.9}$$

$$\geq -\sum_{j \in A_i} \theta_{ij}(x_j - y_i)\Delta t_{ij}^{max} + d_i + \min_{k \in D_i}(y_k + \delta_{ik}) \tag{3.10}$$

The relationship between the bounds and the final switching window are illustrated in Figure 3.3.

3. Fixed Point Computation

Fixed point computation provides a convenient vehicle to explore the underlying properties of how the computation precedes. In this section, we propose the formulation and point out some important properties.

3.1 Formulation

Let $\mathbf{x} \in \mathbf{R}^{2N}$ be a switching window configuration, where N is the number of nets or timing nodes in switching windows computation. For N nets, we need $2N$ variables to represent the latest arrival times, x_i's, and the earliest arrival times, y_i's, respectively (if rise and fall switching windows are considered separately, $4N$ variables are needed). Let $\mathbf{f} : \mathbf{R}^{2N} \to \mathbf{R}^{2N}$ be a mapping or transformation from $\mathbf{x}$ to a new switching window configuration $\mathbf{x}'$ considering the crosstalk noise based on the switching windows' overlapping calculated according to $\mathbf{x}$.

The objective of switching windows computation thus can be formulated as finding a **fixed point**, $\mathbf{x}^*$, such that $\mathbf{x}^* = \mathbf{f}(\mathbf{x}^*)$ [ZSN01]. Specifically, iteration equations are written as

$$x_i^{(n+1)} = \sum_{j \in A_i} \theta_{ij}(x_i^{(n)} - y_j^{(n)})\Delta t_{ij}^{max} + d_i + \max_{k \in D_i}(x_k^{(n)} + \delta_{ik})$$

and

$$y_i^{(n+1)} = -\sum_{j \in A_i} \theta_{ij}(y_i^{(n)} - x_j^{(n)})\Delta t_{ij}^{max} + d_i + \min_{k \in D_i}(y_k^{(n)} + \delta_{ik}).$$

With an initial guess, $\mathbf{x}^{(0)}$, we can perform **fixed point iterations** as

$$\mathbf{x}^{(1)} = \mathbf{f}(\mathbf{x}^{(0)}), \ \mathbf{x}^{(2)} = \mathbf{f}(\mathbf{x}^{(1)}), \ \ldots \ \mathbf{x}^{(n)} = \mathbf{f}(\mathbf{x}^{(n-1)}) \ \ldots \quad (3.11)$$

until it converges. This process is usually referred to as a **fixed point computation** [EMU96].

3.2 Fixed Point Iteration for Switching Windows Computation

Let D be a closed and bounded domain in $\mathbf{R}^{2N}$, and let $\mathbf{f}(\mathbf{x}) \in D$ for all $\mathbf{x} \in D$. For any two points $\mathbf{x_1}, \mathbf{x_2} \in D$, if there exists a constant L, such that

$$||\mathbf{f}(\mathbf{x_1}) - \mathbf{f}(\mathbf{x_2})|| < L||\mathbf{x_1} - \mathbf{x_2}||,$$

where $|| \cdot ||$ denotes a norm for the vector space D, $\mathbf{f}(\mathbf{x})$ is called a *Lipschitz function*, and L is called a *Lipschitz constant*. Using the fixed point computation [EMU96], if $0 \leq L < 1$, the fixed point iteration converges and guarantees a unique convergence point (fixed point), given any initial $\mathbf{x}_0 \in D$. This is a sufficient condition for existence, convergence and uniqueness [EMU96].

In practice, for the one dimensional case, L is roughly estimated as

$$L = \sup \frac{df_i(x_i)}{dx_i} = \sup \sum_{j \in A_i} \theta'_{ij}(x_i - y_j)\Delta t_{ij}^{max}$$

and

$$\Delta t_{ij}^{max} = \Delta V_{max}\frac{\tau_i}{V_{DD}},$$

where sup is an upper bound (maximum value) function, ΔV_{max} is the maximum noise induced by an aggressor and τ_i and τ_j are the slew time of a victim and an aggressor, respectively. In general, L is not bounded by 1.0. It depends on the underlying models, which are discussed in Section 3.4. If a ramp waveform is assumed, $\sup \theta'_{ij}(x_i - y_j)$ can be estimated as $\frac{1}{\tau_j}$, since the victim delay(x_i) starts to change when the aggressor waveform overlaps with the

50% V_{DD} point of the victim waveform, and saturates as 1 when the aggressor ramp doesn't overlap with the 50% V_{DD} point of the victim waveform. The time interval is equal to τ_j, and the slope is therefore $\frac{1}{\tau_j}$.

To give an example $L > 1$, consider the electrical model used in 1.3.2. We can estimate:

$$\Delta V_{max} \approx V_{DD} \frac{C_x R_v}{\tau_j},$$

so we have:

$$L = \theta'_{ij}(x_i - y_j)\Delta t_{ij}^{max} \quad = \quad \frac{1}{\tau_j}\Delta V_{max}\frac{\tau_i}{V_{DD}} \tag{3.12}$$

$$\approx \quad \frac{1}{\tau_j}V_{DD}\frac{C_x R_v}{\tau_j}\frac{\tau_i}{V_{DD}} = \frac{\tau_i C_x R_v}{\tau_j^2} \tag{3.13}$$

It is not difficult to find real design values that make the L above greater than 1. Consider an example circuit in Figure 3.4. Suppose the aggressor net G is driven by a very strong AND gate, so its switching window is not affected by the weak aggressor net H. We can calculate $\Delta t_{H,G}^{max}$ and x_H as:

$$\Delta t_{H,G}^{max} \approx \frac{0.05pF \times 400ohm \times 2ns}{0.2ns} = 0.2ns,$$

and

$$x_H = f(x_H) \quad = \quad \theta_{H,G}(x_H - y_G)\Delta t_{H,G}^{max} + x_D + \delta_{H,D} \tag{3.14}$$

$$= \quad \theta(x_H) \times 0.2ns + 1.9ns + 1ns, \tag{3.15}$$

where $\theta(x_H)$ is shown in Figure 3.4. If $x_H^{(0)}$ starts from 2.9ns, we can get the final x_H^* as 2.9ns. If $x_H^{(0)}$ starts from $+\infty$, we can get the final x_H^* as 3.1ns. This shows that we can have multiple convergence points depending on the initial switching window configuration.

For a one dimensional iteration function $f(x)$ in Figure 3.5, we can see:

$$f(x) = \sum_{j \in A_i} \theta_{ij}(x - y_j)\Delta t_{ij}^{max} + d_i + \max_{k \in D_i}(x_k + \delta_{ik}), \tag{3.16}$$

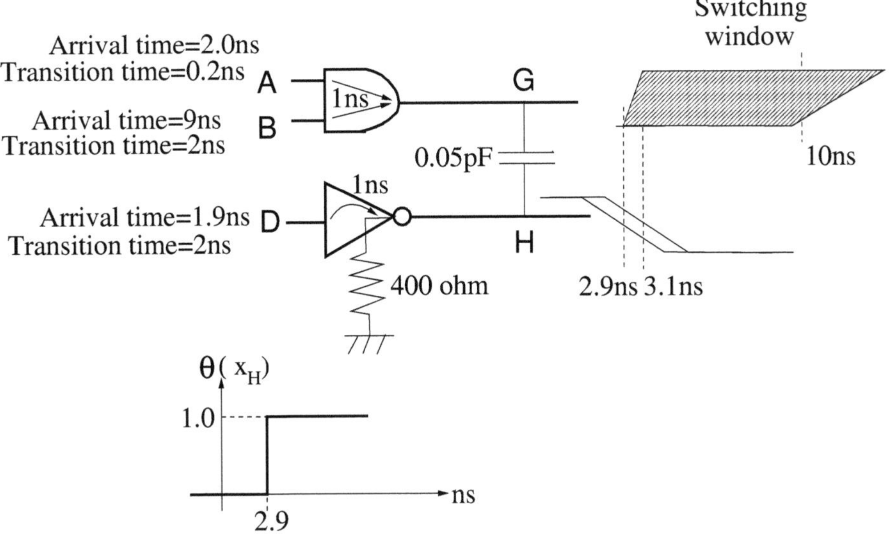

Figure 3.4. A circuit example for multiple convergence points

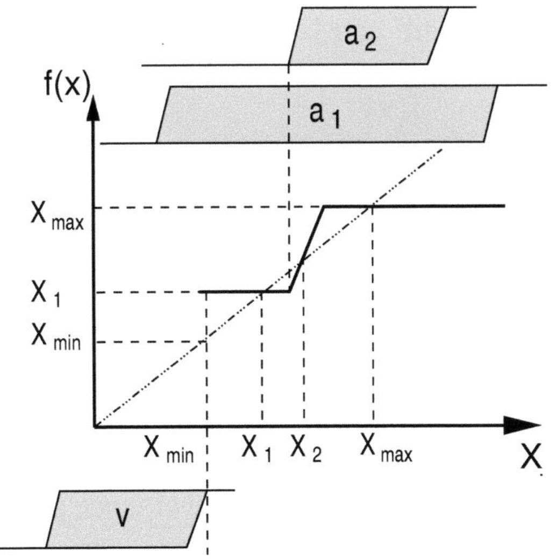

Figure 3.5. Realizable arrival time function

where x^{min} and x^{max} are the lower and the upper bounds described in Section 3.2.1, and victim net v and two aggressors a_1 and a_2 together create function $f(x)$, where x_1, x_2 and x_{max} are the fixed points. A repeated substitution procedure that replaces the argument with its output value can be used to converge the $x^{(n)}$ sequence. For the one dimensional case and a continuous function, a sufficient condition for convergence is given as $|f'(x)| < 1$. For $\mathbf{f}(\mathbf{x})$, the sufficient condition [EMU96] is $||\mathbf{J}|| < 1$, where $\mathbf{J}$ is the Jacobian matrix of $\mathbf{f}(\mathbf{x})$.

3.3   Multiple Convergence Points and Unstable Fixed Point

Since L is not generally bounded by 1.0, it is easy to produce multiple convergence points in switching windows computation process. The actual convergence point depends on the initial switching windows [ZSN01].

For fixed point computation, a unique convergence point requires a *Lipschitz constant L* less than 1.0. In switching windows computation, L is not bounded for discrete models introduced in Section 3.4, and therefore there can be multiple fixed points. Even for a

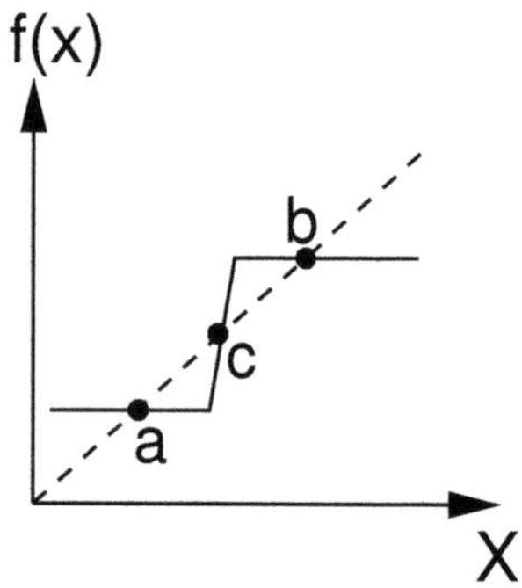

Figure 3.6.   Multiple convergence points

continuous model, it is possible for L to be greater than 1.0. For example, in Figure 3.6, points a, b and c are all fixed points, and the initial condition determines which fixed point the iteration converges to. Notice that point c is **unstable** because $f'(x)|_{x=c} > 1$. A small perturbation can drive convergence toward points a or b. Unstable fixed points cannot be obtained through fixed point computation.

3.4 Tightening Bounds

A conventional convergence scenario uses infinite switching windows for the initial condition to include all the noise effects in the beginning and shrinks the switching windows in the subsequent iterations [ARP00, XCMS00]. [ZSN01] shows that starting from the infinite switching windows, the process can converge to a looser upper bound of switching windows.

Using Eq. (3.1) and (3.2), we can prove this monotonicity by induction as follows.

THEOREM 3.1 *If in the initial two steps, $x_i^{(0)} \geq x_i^{(1)}$ and $y_i^{(0)} \leq y_i^{(1)}$, for all $i = 1..N$, then $x_i^{(n)}$ forms a monotonically non-increasing sequence, and $y_i^{(n)}$ forms a monotonically non-decreasing sequence.*

Proof: The proof is by induction on the iteration number. The base case is clearly

$$x_i^{(0)} \geq x_i^{(1)} \quad \text{and} \quad y_i^{(0)} \leq y_i^{(1)}, \quad \text{for} \quad i = 1..N.$$

Assume that

$$x_i^{(k-1)} \geq x_i^{(k)} \quad \text{and} \quad y_j^{(k-1)} \leq y_j^{(k)} \quad \text{for} \quad n \geq k \geq 1,$$

so

$$x_i^{(n)} - y_j^{(n)} \leq x_i^{(n-1)} - y_j^{(n-1)}.$$

θ_{ij} is a non-decreasing function, so we have

$$\theta_{ij}(x_i^{(n)} - y_j^{(n)}) \leq \theta_{ij}(x_i^{(n-1)} - y_j^{(n-1)}).$$

Also

$$\Delta t_{ij}^{max} > 0 \quad \text{and} \quad \max_{k \in D_i}(x_k^{(n)} + \delta_{ik}) \leq \max_{k \in D_i}(x_k^{(n-1)} + \delta_{ik}).$$

Therefore,

$$x_i^{(n+1)} = \sum_{j \in A_i} \theta_{ij}(x_i^{(n)} - y_j^{(n)})\Delta t_{ij}^{max} + d_i + \max_{k \in D_i}(x_k^{(n)} + \delta_{ik})$$

$$\leq \sum_{j \in A_i} \theta_{ij}(x_i^{(n-1)} - y_j^{(n-1)})\Delta t_{ij}^{max} + d_i + \max_{k \in D_i}(x_k^{(n-1)} + \delta_{ik}) = x_i^{(n)}.$$

Similarly, the $y_i^{(n)}$ sequence can be proved symmetrically. $\square$

Corollary 1.1 *If the initial configuration starts from the maximum switching windows, $x_i^{(n)}$ forms a monotonically non-increasing sequence, and $y_i^{(n)}$ forms a monotonically non-decreasing sequence.* This result actually shows that the switching window shrinks starting from the maximum switching window – that is, the upper bound is reduced in each iteration. Thus, the accuracy of the noisy switching windows' bound computation depends on how much run time can be afforded. The results are still valid upper bounds of switching windows even before convergence. This is a strictly conservative process.

Similarly, using the minimum switching windows as the initial condition, the lower bound increases as the convergence process precedes. We have the following theorem and the corollary.

THEOREM 3.2 *If in the initial two steps, $x_i^{(0)} \leq x_i^{(1)}$ and $y_i^{(0)} \geq y_i^{(1)}$ for any $i = 1..N$, $x_i^{(n)}$ forms a monotonically non-decreasing sequence, and $y_i^{(n)}$ forms a monotonically non-increasing sequence.*

Proof: The proof is by induction on the iteration number. The base case is clearly

$$x_i^{(0)} \leq x_i^{(1)} \quad \text{and} \quad y_i^{(0)} \geq y_i^{(1)}, \quad \text{for} \quad i = 1..N.$$

Assume that

$$x_i^{(k-1)} \leq x_i^{(k)} \quad \text{and} \quad y_j^{(k-1)} \geq y_j^{(k)} \quad \text{for} \quad n \geq k \geq 1,$$

so

$$x_i^{(n)} - y_j^{(n)} \geq x_i^{(n-1)} - y_j^{(n-1)}.$$

θ_{ij} is a non-decreasing function, so we have

$$\theta_{ij}(x_i^{(n)} - y_j^{(n)}) \geq \theta_{ij}(x_i^{(n-1)} - y_j^{(n-1)}).$$

Also

$$\Delta t_{ij}^{max} > 0 \quad \text{and} \quad \max_{k \in D_i}(x_k^{(n)} + \delta_{ik}) \geq \max_{k \in D_i}(x_k^{(n-1)} + \delta_{ik}).$$

Therefore,

$$x_i^{(n+1)} = \sum_{j \in A_i} \theta_{ij}(x_i^{(n)} - y_j^{(n)})\Delta t_{ij}^{max} + d_i + \max_{k \in D_i}(x_k^{(n)} + \delta_{ik})$$

$$\geq \sum_{j \in A_i} \theta_{ij}(x_i^{(n-1)} - y_j^{(n-1)})\Delta t_{ij}^{max} + d_i + \max_{k \in D_i}(x_k^{(n-1)} + \delta_{ik}) = x_i^{(n)}.$$

Similarly, the $y_i^{(n)}$ sequence can be proved symmetrically. $\square$

Corollary 2.1 *If the initial configuration starts from the noise-less(smallest) switching windows, $x_i^{(n)}$ forms a monotonically non-decreasing sequence, and $y_i^{(n)}$ forms a monotonically non-increasing sequence.*

This corollary actually provides a method to obtain the tightest switching windows defined in Section 3.2.

These results create a monotonic transformation during fixed point iteration suggested by [ZSN01].

4. Coupling Models

In this section, we consider the underlying models for calculating noise. Discrete models are easier and faster to calculate and, in general, give a bound for crosstalk noise. However, the error bound can be far off from the correct noise bound computed using a continuous model.

Crosstalk noise induces a voltage glitch on a victim and causes a timing variation. The amount of the delta delay in a timing calculation can be determined by aligning the noise peak with the victim waveform so that the superimposed waveform peak reaches the switching threshold (usually 50% of power rail voltage) [GRP98]. Figure 3.7 shows this method. If the victim's waveform is simplified as an ideal ramp with a slew time τ_v, the maximum delta delay Δt^{max} can be written as

$$\Delta t^{\mathrm{max}} = \tau_v \frac{\Delta V_{max}}{V_{DD}}.$$

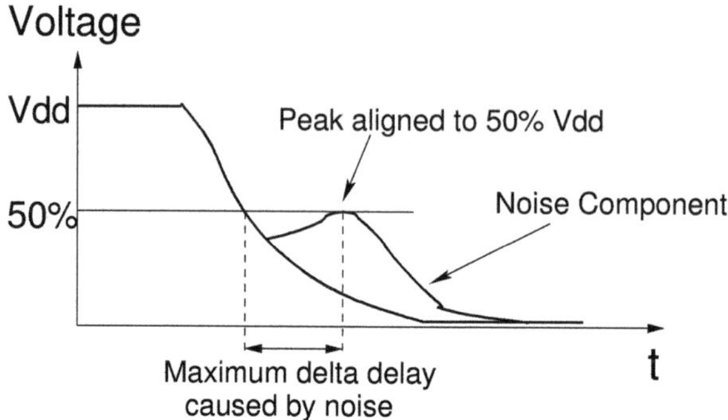

Figure 3.7.   Determination of the worst delta delay caused by crosstalk noise

4.1   Noise Calculation Model

Traditional STA often ignores all the coupling effects and replaces any coupling capacitance with a grounded one. Conventionally, to calculate the coupling delay on each interconnect, a **discrete coupling factor model** uses 1X grounded capacitance when the neighboring net is quiet, 2X for the opposite direction switching, and 0 for the same direction switching [Sap00, ZSN01]. Determining which factor to use depends on two nets' switching time and the directions. However, Chapter 2 has been shown that the coupling noise can result in as much as a 3X capacitance effect when calculating the coupling delay [KMS00, CKK00a]. Besides, a discrete coupling factor model has a discontinuity between the boundaries when a coupling factor is changed to another factor. On the contrary, a **continuous coupling factor model** can be used to avoid this discontinuity on the boundaries and increase accuracy [KMS00, CKK00a]. Chapter 2 has a detailed discussion of how to determine this continuous coupling factor. However, the convergence rate of discrete coupling factor model is faster than a continuous one in practice.

More accurate models have been proposed using superposition to calculate the total crosstalk noise without using any coupling factor or decoupling the coupling capacitance, such as [ARP00, LBB$^+$00, TCE00].

In [CKK00b] and Chapter 4, we propose a model, in which instead of direct substitution of x_i in Eq. (3.1) to evaluate the crosstalk noise, they decompose an arrival time into a component that is contributed from the driving gate as

$$\overline{x}_i = d_i + \max_{k \in D_i}(x_k + \delta_{ik}) \quad \text{and} \quad \overline{y}_i = d_i + \min_{k \in D_i}(y_k + \delta_{ik}),$$

and, based on these arrival times $\overline{x}_i$ and $\overline{y}_i$, evaluate the noisy arrival times as

$$x_i = \sum_{j \in A_i} \theta_{ij}(\overline{x}_i - \overline{y}_j)\Delta t_{ij}^{max} + \overline{x}_i$$

and

$$y_i = -\sum_{j \in A_i} \theta_{ij}(\overline{x}_j - \overline{y}_i)\Delta t_{ij}^{max} + \overline{y}_i.$$

It avoids pessimistically taking the noisy arrival time into account for switching windows' overlapping. This model can still be calculated using fixed point computation.

4.2 Switching Windows Overlapping Model

For multiple aggressors, the worst case noise should be calculated based on the switching window constraints – that is, due to the path delays, the aggressors' switching window might not be able to align arbitrarily to create the maximum noise.

In [SNEZ97c], the authors have proposed a mixed integer programming technique to find the worst possible noise. This model assumes "sharpness" of the noise peaks and it is considered a **discrete overlapping model**, which only allows either a complete noise contribution or zero noise. That is to say, $\theta(\cdot)$ is a step function in Eq. (3.1) and Eq. (3.2). Figure 3.8 shows a noise function in such models. On the contrary, a **continuous overlapping model** allows the noise contribution to be fractional according to the noise glitch waveform on the victim net and the overlapping range. Figure 3.1 is an example of $\theta(\cdot)$ representing a continuous model. Many efficient methods have been proposed to find the maximum noise or the maximum delta delay using these types of models. For example, [TCE00, CKK00b] have proposed envelope waveform methods, [XMS00b, CMS01] formulate it as a weighted channel density problem and give an algorithm with $O(MlogM)$ complexity, where M is the number of aggressors.

Note that if an infinite slope is assumed on the boundary of switching windows, a continuous overlapping model becomes a discrete overlapping model.

4.3   Discontinuity in Discrete Models

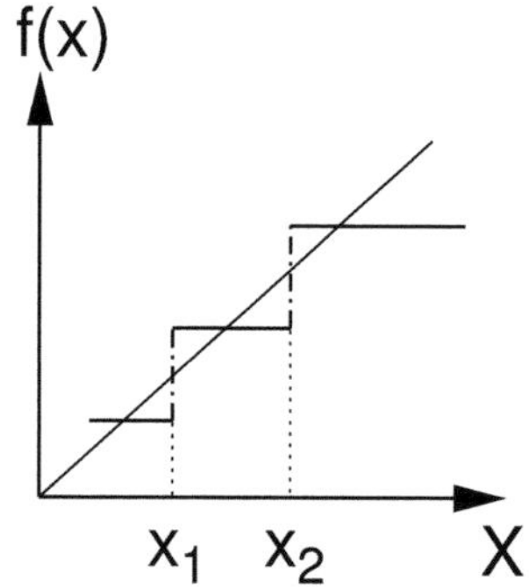

Figure 3.8.   Discontinuity in noise $f(x)$ when using a discrete model

All of the discrete models mentioned above suffer from a drawback of discontinuity on the boundary when a discrete factor is changed to another discrete factor, or when the overlapping condition is changed. The noise is discontinuous when increasing or decreasing gate delays. Figure 3.8 shows an example of discontinuity at x_1 and x_2 when using a discrete overlapping model.

If the discrete model is designed carefully, it can be an upper bound of crosstalk noise, which is considered to be very useful in STA. The example in Figure 3.9 shows that the discrete model converges at point d while the continuous model converges at point c, which is bounded by point d.

4.4 Error Bound between Discrete and Continuous Models

After convergence of switching windows, suppose no error was incurred in the previous stage delay for a single pair of coupling nets. The error incurred due to the use of a discrete model can be as large as $\Delta t^{\max}$, the maximum delta delay between a victim and an

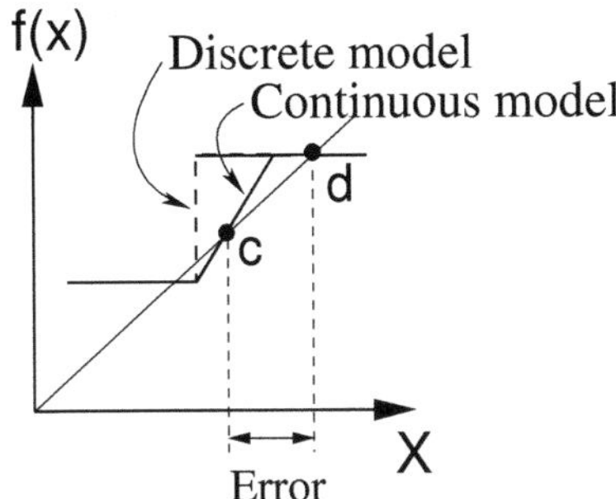

Figure 3.9.   Error incurred due to a discrete model

aggressor net pair given the same initial configuration of switching windows. The error can be written as:

$$\epsilon_{ij} = (\theta_{ij}^{con}(x_i - y_j) - \theta_{ij}^{dis}(x_i - y_j))\Delta t_{ij}^{\max} \leq \Delta t_{ij}^{\max},$$

where $\theta_{ij}^{con}(\cdot)$ is the continuous overlapping function, and $\theta_{ij}^{dis}(\cdot)$ is the discrete overlapping function. Figure 3.9 shows an example where the discrete model converges to point d and the continuous model converges to point c. If every aggressor aligns exactly at the same time, the error bound can be as large as the sum of the maximum delta delays:

$$\epsilon_i^{net} \leq \sum_{j \in A_i} \theta_{ij}(x_i - y_j)\Delta t_{ij}^{\max} \leq \sum_{j \in A_i} \Delta t_{ij}^{\max}.$$

Moreover, the error can propagate forward along a timing path and accumulate to include all the maximum delta delays induced from all the aggressors. Therefore, the error bound for a noisy longest path delay is equal to the difference between the upper bound and the lower bound of the ending net's switching window of the path:

$$\epsilon(x_i) \le x_i^{\max} - x_i^{\min} = \sum_{j \in A_i} \Delta t_{ij}^{\max} + \max_{k \in D_i}(x_k^{\max} + \delta_{ik}) - \max_{k \in D_i}(x_k^{\min} + \delta_{ik}),$$

and similarly, the error bound for a noisy shortest path delay is equal to:

$$\epsilon(y_i) \le y_i^{\max} - y_i^{\min} = \sum_{j \in A_i} \Delta t_{ij}^{\max} + \min_{k \in D_i}(y_k^{\max} + \delta_{ik}) - \min_{k \in D_i}(y_k^{\min} + \delta_{ik}),$$

4.5   Non-Monotone Property

Ordinarily, when delay of a gate in a circuit decreases, the longest delay of the circuit will also decrease or remain the same. In the case of crosstalk, the noisy longest delay of a circuit might increase. For example, consider a case in Figure 3.10, where if the arrival time of switching window a is reduced, net a might start to attack net v due to the overlapping of switching windows, resulting in a delay increase in net v. In terms of Eq. (3.1), as y_j decreases,

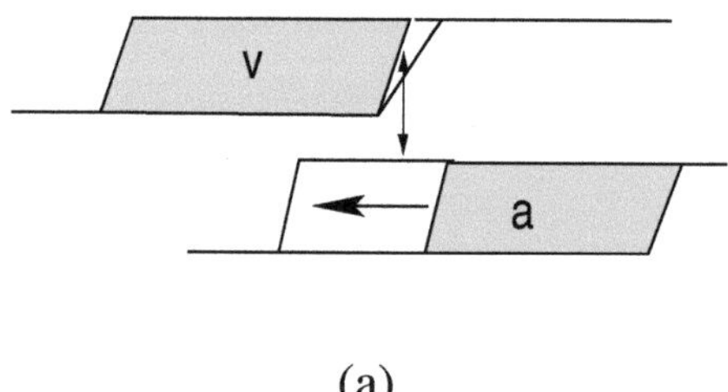

(a)

Figure 3.10.   Reducing a gate delay resulting in a longer path delay

$\theta_{ij}(x_i - y_j)$ increases, resulting in an increase of x_i on the left hand side. Using just one operational condition to determine x_i and y_j can result in an optimistic evaluation of crosstalk noise.

Using a floating delay model [DKM93] (e.g. see Figure 3.11), which assumes zero for any earliest arrival time, i.e. $y_j = 0$ in Eq. (3.1), it is still impossible to solve this problem. If we set $y_j = 0$, Eq. (3.1) becomes

$$x_i = \sum_{j \in A_i} \theta_{ij}(x_i)\Delta t_{ij}^{max} + d_i + \max_{k \in D_i}(x_k + \delta_{ik}) \qquad (3.17)$$

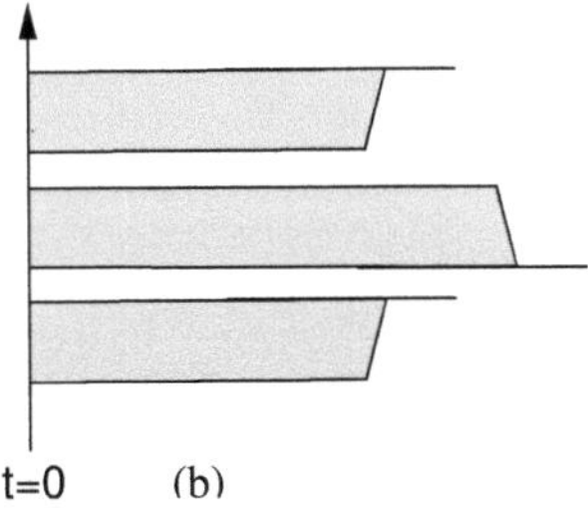

Figure 3.11.   Floating mode delay model

$$= \sum_{j \in A_i} \Delta t_{ij}^{max} + d_i + \max_{k \in D_i}(x_k + \delta_{ik}) = x_i^{max}, \quad (3.18)$$

which does not provide any useful information about switching windows.

5. Convergence of Switching Windows Computation

In this section, we argue the convergence of switching windows computation.

5.1 Proof of Convergence

THEOREM 3.3 *The iteration in Eq. (3.11) converges to a fixed point* $\mathbf{x}^* = \mathbf{f}(\mathbf{x}^*)$, *given the initial switching window configuration as* $\mathbf{x}^{(0)} = \mathbf{x}^{min}$ *or* $\mathbf{x}^{(0)} = \mathbf{x}^{max}$

Proof: Convergence of the iteration can be proved by the following facts [ARP00, ZSN01]:

1 x_i's and y_i's have an upper and a lower bounds, as shown in Section 3.2.1.

2 If starting from $\mathbf{x}^{(0)} = \mathbf{x}^{min}$, by Corollary 2.1, $x_i^{(n)}$ forms a non-decreasing sequence and $y_i^{(n)}$ forms a non-increasing sequence. The iteration converges, since the sequences are bounded.

3 Using Corollary 1.1, we can prove similarly for the initial condition starting from $\mathbf{x}^{(0)} = \mathbf{x}^{\max}$. $\square$

Some switching-window overlapping models might not have monotonicity for $\mathbf{f}(\mathbf{x})$ with respect to x_i. For example, when an aggressor's latest arrival time is much less than a victim's arrival time, the switching window of the victim is not affected. This effect can be captured by adding an extra term $-\theta_{ij}(x_i - x_j)$ to Eq. (3.1) as:

$$x_i = \sum_{j \in A_i} (\theta_{ij}(x_i - y_j) - \theta_{ij}(x_i - x_j))\Delta t_{ij}^{max} + d_i + \max_{k \in D_i}(x_k + \delta_{ik})$$

(3.19)

However, as shown in Figure 3.12, Eq. (3.19) has a decreasing portion, leading to oscillation among points a, b, c and d, so the iteration cannot converge. To remedy the decreasing portion, the aggressor's

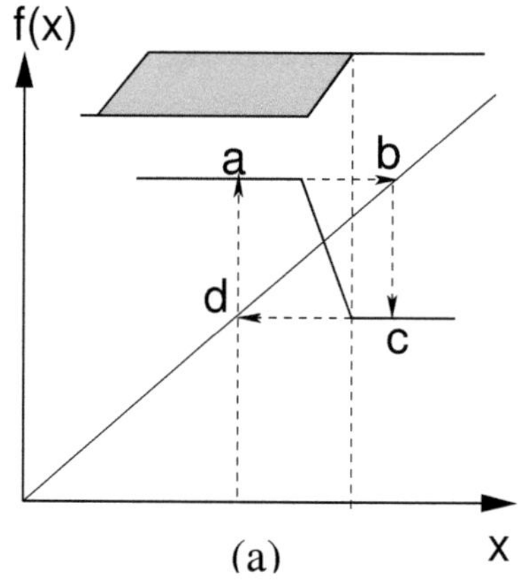

Figure 3.12.   A decreasing portion resulting in non-convergence

switching window must be extended to infinity, as shown in Figure 3.13, when calculating crosstalk noise between coupling nets, and hence Eq. (3.1) is mostly used. Without the decreasing portion, Eq. (3.1) is monotonically increasing for x_i and x_k, and monotonically decreasing for y_j. Theorem 3 equivalently shows there is at least one fixed point in D. The next step is to see if if there is some looping structure in Eq. (3.1) (i.e. oscillation among points in the

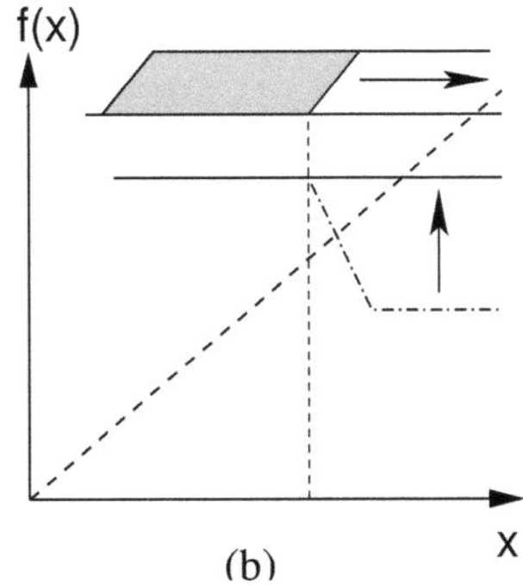

Figure 3.13. Extending the aggressor's switching window to infinity

iteration):

$$\mathbf{x}^{(n+k)} = \mathbf{f}^{(n)}(\mathbf{x}^{(k)}) \quad \text{and} \quad n > 1.$$

THEOREM 3.4 *Given any initial conditions with* $x_i^{\min} \le x_i^{(0)} \le x_i^{\max}$ *and* $y_i^{\min} \le y_i^{(0)} \le y_i^{\max}$*, the iteration in Eq. (3.11) converges.*

Proof sketch: Without loss of generality, we assume x_i affects y_j and y_j affects x_k, and x_k affects x_i in Eq. (3.1). Any increase in x_i leads to decrease in y_j, and subsequently increase in x_k, which in turns increases x_i. This excludes the possibility of oscillation in the iteration – that is, the iteration converges given any initial conditions with $x_i^{\min} \le x_i^{(0)} \le x_i^{\max}$ and $y_i^{\min} \le y_i^{(0)} \le y_i^{\max}$. $\quad\square$

5.2 Computational Complexity

For a discrete overlapping model, we have at most $O(NM) \ge O(L)$ coupling edges, where N is the number of nets, M is the maximum number of aggressor nets for any victim net, and L is the total number of the coupling edges in a circuit. In each pass of calculation of $\mathbf{f}(\mathbf{x})$, at least one coupling edge's state is finalized such that the edge either contributes noise to its victim or not for all the subsequent passes. If no coupling edge changes its state, there is no update to the noisy switching windows, so the iteration converges. For each pass, we need to examine $O(N)$ nets against their $O(M)$ aggressors to identify switching windows that overlap. Thus, the total complexity is $O(N^2 M^2) \geq O(NML)$ or . In [Sap00], an algorithm with complexity $O(N^2 M)$ is suggested without counting the cost of checking against $O(M)$ aggressors for detecting overlaps on each net. In practice, this algorithm converges quite fast within 3 to 5 iterations.

If an event-driven style of computation is used [CKK00b] (or Chapter 4), the total complexity is still the same because there are $O(N)$ nets in an STA timing graph to update, and for each update to a switching window, we need to check it against $O(M)$ aggressors to detect overlapping and trigger new update events. There are $O(L)$ coupling edges, thus we need $O(NML)$ operations for an event-driven style of computation. They conclude with an efficient algorithm similar to the approach of using the Gauss-Seidel method.

5.3   Convergence Rate

For fixed point computation under a continuous overlapping model, the convergence rate must be considered. For some local regions, there is a local Lipschitz constant L, such that $L < 1$, and the fixed point iteration converges. The convergence rate is determined by this local Lipschitz constant L. Provided $\mathbf{x}^{(n)}$ is close enough to $\mathbf{x}^*$, the error is bounded by [EMU96]:

$$||\mathbf{x}^{(n)} - \mathbf{x}^*|| \leq \frac{L^n}{1-L}||\mathbf{x}^{(1)} - \mathbf{x}^{(0)}||$$

or

$$||\mathbf{x}^{(n)} - \mathbf{x}^*|| \leq \frac{L}{1-L}||\mathbf{x}^{(n)} - \mathbf{x}^{(n-1)}||.$$

Since L can be greater than 1.0 in some local areas (see Section 3.3.2), it could have some local divergent sequences as shown in Figure 3.14 for one dimensional case. However, the ending game of convergence is still dominated by the convergent L value, which is closest to 1.0.

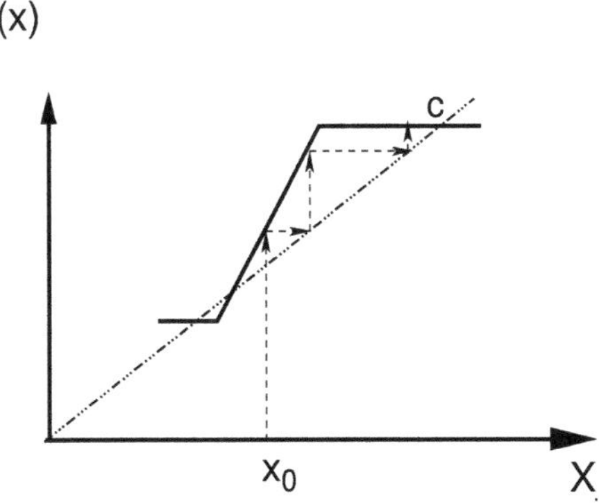

Figure 3.14.   Local divergence

5.4   Least Evaluation of Coupling RC Networks

In practice, the computation of coupling RC networks is the performance bottleneck for switching windows computation. Detailed RC extraction can generate an RC network for a net with over thousands of nodes. Efficient RC reduction and efficient coupling noise calculations are required to speed up the calculation. However, an initial switching window configuration also affects the number of RC calculations. If each switching window starts from its noiseless (nominal) switching window, and the iteration process changes the coupling edges' state one-by-one as switching windows increase, no coupling RC network calculations performed during iteration for noise will be wasted. In case of the initial condition starting from the maximum noisy switching windows, all of the coupling RC networks must be computed for the maximum delta delays. However, some of coupling edges might change its state from active to inactive in the process of iteration. The coupling RC network calculations for these edges are wasted compared with the noiseless initial condition.

5.5   Speed-up of Convergence

A number of speed-up methods for convergence have been proposed [XCMS00, CKK00b, Sap00, ZSN01]. Most of the techniques are similar to Gauss-Jordan, Gauss-Seidel, or update (event)-driven calculations. A Gauss-Seidel style calculation for fixed point computation uses any updated information as soon as available. For example, the iteration function can be modified as

$$x_i^{(n+1)} = \sum_{j \in A_i} \theta_{ij}(x_i^{(n)} - y_j^{(n+1)})\Delta t_{ij}^{max} + d_i + \max_{k \in D_i}(x_k^{(n+1)} + \delta_{ik}).$$

Usually, $y_j^{(n)}$ is used if $y_j^{(n+1)}$ is not available at the moment when calculating $x^{(n+1)}$. Techniques are thus focused on how to maximize the use of $y_j^{(n+1)}$ instead of $y_j^{(n)}$, and the use of $x_k^{(n+1)}$ instead of $x_k^{(n)}$.

Another speed-up method is to replace the fixed point computation by a Newton-Raphson iteration, which has a quadratic convergence rate. However, computing each Jacobian matrix element needs $O(N)$ operations, and the total complexity to build the Jacobian matrix is $O(N^3)$. Inverting the Jacobian matrix might need as many as $O(N^3)$ operations besides the original computation cost of $\mathbf{f}(\mathbf{x})$. Moreover, singularity problem should be handled with care. Aitken Δ^2 method [EMU96] can be applied to each arrival time to quickly achieve local convergence. But neither of the two methods guarantees finding the tightest noisy switching windows.

6.   Conclusion

Switching windows computation can be well-controlled by careful selection of the underlying models. In this chapter, we show, formulate, and prove the various numerical properties from a numerical fixed point computation perspective. These could serve as a theoretical foundation for switching windows computation.

Chapter 4

SPEEDING-UP SWITCHING WINDOW COMPUTATION

After spatial and electrical pruning, temporal pruning may be applied. The key element of temporal pruning is to compute switching windows, which are an important technique to accurately calculate crosstalk effects as discussed in Chapter 3. In this chapter, we present and compare multiple scheduling algorithms to compute switching windows for static timing analysis in the presence of crosstalk noise. We also introduce an efficient technique to evaluate the worst case alignment of multiple aggressors.

1. Introduction

Static timing analysis has been studied for more than a decade [Sas93, DKMW94, DKM93]; however, these studies have not involved any crosstalk coupling analysis. Recently, [She98b] provides a design methodology to avoid coupling noise and addresses static analysis of noise at the transistor level. [Kir97] analyzes the functional aspect of how signals couple together. In [CK99], a formulation is proposed to calculate the maximum noise, but it only applies to small circuits due its complexity. [GRP98] proposes an algorithm to calculate the worst case aggressor alignment due to coupling. [CKK00c] shows that the Miller factor upper bound is

3X instead of 2X for the maximum coupling delay. For a design in DSM, the functional aspect of crosstalk coupling is almost impractical to analyze. STA with crosstalk coupling effects serves a very practical and efficient way to verify a circuit design that will not violate any timing constraints [FFP+00, TCE00].

In this chapter, we address the problem of static timing analysis by considering coupling effects. Unlike traditional STA, the critical path delay cannot be obtained simply by topological traversal and functional analysis of false paths. **Switching windows**, within which a node makes transitions, are key to determining whether the coupling noise can affect timing. Only when two coupling nodes have overlapping switching windows can their timing change due to their coupling. However, switching windows depend on the signal timing itself, so there is a circular problem to resolve. Therefore, we propose an event-driven algorithm to solve this mutual dependency problem, resolving cycles through causality. We assume a single worst case driver resistance and apply superposition of waveforms extensively. This is a conservative assumption. The result is an upper bound of the actual switching window.

This chapter is organized as follows: First, we review necessary background and definitions. In Section 4.3, we discuss the alignment of multiple aggressors for worst-case delay. In Section 4.4, we present an event-driven algorithm in detail, including proof, event scheduling techniques, complexity analysis, and efficiency issues. In Section 4.5, we show some experimental results.

2.   Background and Definitions

Recall that for a pair of coupling nodes, a node which suffers from the coupling noise is called a **victim node**, and other nodes that contribute the noise are called **aggressor nodes**. The **worst case delay** of a node(Figure 4.1) is the minimum or maximum delay considering all topological minimum or maximum delay dependencies, and the worst case crosstalk coupling. For example,

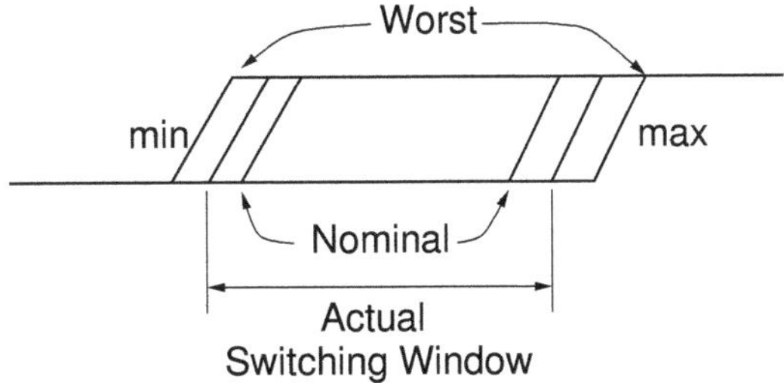

Figure 4.1. Min/Max Timing

the worst case timing can be computed using zero coupling capacitance for min delay and 3X coupling capacitance for max delay, respectively. The **nominal delay** of a node is defined as the delay calculated when each aggressor is quiet – that is, using 1X coupling capacitance for delay calculation. The worst case switching window (Figure 4.1) thus forms the outer bound of the actual switching window, and the nominal case switching window forms the inner bound.

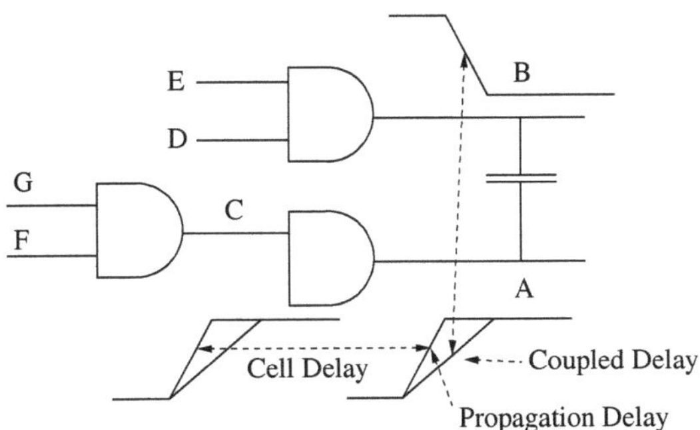

Figure 4.2. Propagation/Coupled Delay

The **coupled delay** of a node (Figure 4.2) is the delay due to the aggressors' coupling of the node. It is computed from the aggressor's maximum coupling noise to achieve the min or max delay. The **propagation delay** of a node (Figure 4.2) is the delay due to the previous stage delay. It is computed from the previous stage coupled delay plus the cell delay.

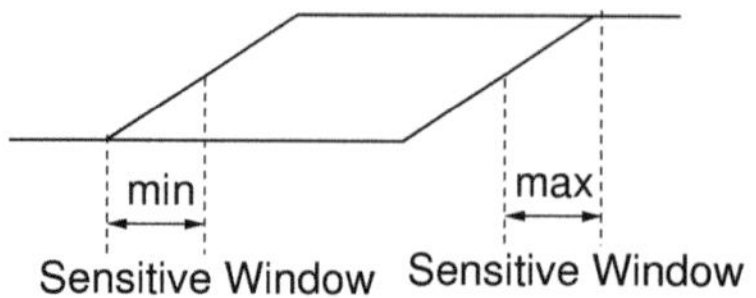

Figure 4.3.   Sensitive Min/Max Windows

Given a node i, the min propagation delay is

$$t_i^{ppg,min} = \min_{j \in FI(i)} \{d_{j,i}^{min} + t_j^{cpl,min}\} \tag{4.1}$$

and the max propagation delay is

$$t_i^{ppg,max} = \max_{j \in FI(i)} \{d_{j,i}^{max} + t_j^{cpl,max}\} \tag{4.2}$$

where $t_i^{ppg,min}(t_i^{ppg,max})$ denotes the min(max) propagation delay, $FI(i)$ is the set of fanins of node i, $d_{j,i}^{min}(d_{j,i}^{max})$ is the min(max) node-to-node delay, and $t_j^{cpl,min}(t_j^{cpl,max})$ denotes the min(max) coupled delay of node j.

Suppose the driving resistance of the aggressor is linear. Coupled delay can be calculated as a superposition of the victim waveform and the coupling noise waveform from the aggressor nodes.

The **min sensitive window** (Figure 4.3) of a node for the min delay is from the rising point to the threshold point of the transition. In this period, the coupling noise can speed up the transition and the delay might be reduced. This period is used to determine the min delay variation because the possible range for a signal to be sped up is just within this window. Similarly, the **max sensitive window** of a node for the max delay is from the threshold point to the end point of the transition. In this period, the coupling noise might slow down the transition and the delay can be increased.

2.1   Piecewise Linear Waveform

For ease of calculation, we assume piecewise linear waveforms. The number of linear segments can be used to trade accuracy and run time for modeling waveforms from circuit level characterization. Moreover, they can be manipulated to do waveform superposition and compute an envelope waveform, as described in Section 4.4. All of the computational complexity is proportional to the number of linear segments in the waveforms.

3. Multiple Aggressor Alignment Problem

In this section, we will discuss how to determine the worst case alignment given multiple aggressor waveforms and a victim waveform. The problem is to align aggressor waveforms to get the maximum or the minimum delay on the victim node. Because each node has a switching window, it restricts the time for a node to make transitions. These switching window constraints restrict the range where a worst case alignment can occur to the ranges where these windows line up. In [GRP98], the authors prove that the worst case delay for a pair of aggressor and victim nodes is when the peak noise of the victim aligns up to the switching threshold of the victim's transition waveform. Based on their result, we address the case where multiple aggressors are aligned, which is common in any STA scenario. We propose an envelope waveform to perform this computation, something relatively easy to compute where the complexity is simply proportional to the number of linear segments in all of the waveforms.

In a physical layout, there could be several aggressors coupling to a victim node, and each of these is constrained by some switching window.

Consider the waveforms of two nodes switching in opposite directions, as shown in Figure 4.4. The problem is equivalent to sliding or convolving the aggressor waveform subject to the switching window constraint to achieve the maximum delay on the victim node. Specifically, we have to find the scenario that maximizes the t_x point in Figure 4.4 [GRP98]. At this point, the waveform of the victim touches the threshold point of the next stage logic gate, mak-

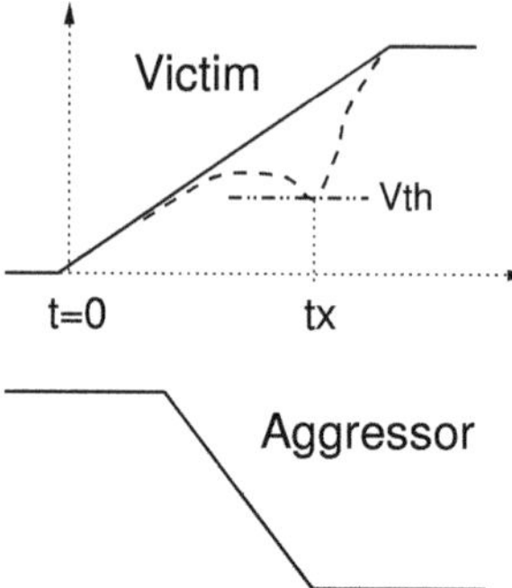

Figure 4.4. Maximum Delay under Coupling Effect

ing a sharp transition between logic states. If the coupling noise waveform continuously slides from the left bound of the aggressor switching window to the right bound, as shown in Figure 4.5, this waveform envelope forms a range and magnitude of noise that could possibly affect the victim waveform. After superposition of this envelope and the victim waveform, the resulting waveform is the worst case waveform envelope of the victim node. The worst case delay can be found on the last point crossing the switching threshold (usually 0.5 Vdd). The bold lines in Figure 4.5 show the noise peak and the corresponding aggressor transition to create this worst case timing.

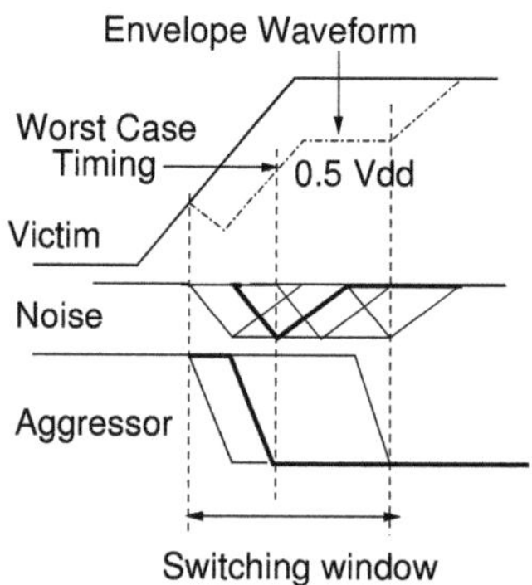

Figure 4.5. Sliding Noise and Envelope Waveform

THEOREM 4.1 *The technique described above can find the worst case alignment which creates the worst case delay on a victim node, given the switching window constraints of multiple aggressors.*

(Proof sketch:) The victim envelope waveform actually depicts the minimum voltage values that the victim waveform can possibly reach over time. Due to the superposition assumption, we can superpose each aggressor envelope waveform on the victim envelope waveform one by one. The resulting waveform envelope is the final worst case voltage that can be reached over time. By tracing this envelope waveform, the worst case delay can be obtained. $\square$

4. Coupling Delay Computation in Presence of Crosstalk Noise

In today's technology, RC delay calculation consumes a major portion of the total computation time for delay calculation. Typical RC delay calculation algorithm involves effective capacitance computation [QPP94] and model order reduction of the RC interconnect [SNEZ97a]. Cell delay computation is a relatively simple computation often via a table lookup. Waveform superposition is another complexity that adds to the whole coupling computation. Therefore, our algorithm is optimized toward reducing the number of coupling computations.

4.1 Algorithm

There are two types of events in our event-driving algorithm. A **coupling event** is the event triggering calculation of the coupling waveform envelope based on the victim and aggressors' waveforms to derive the coupled delay. A **driving event** is the event triggering calculation of the propagation delay based on the previous stages' coupled delays.

Given a circuit with the coupling noise for each victim and aggressor pair, and the waveform that has been characterized, we propose the following event-driven algorithm to compute effective circuit delay: **Coupling Delay Calculation by Event-Driven**

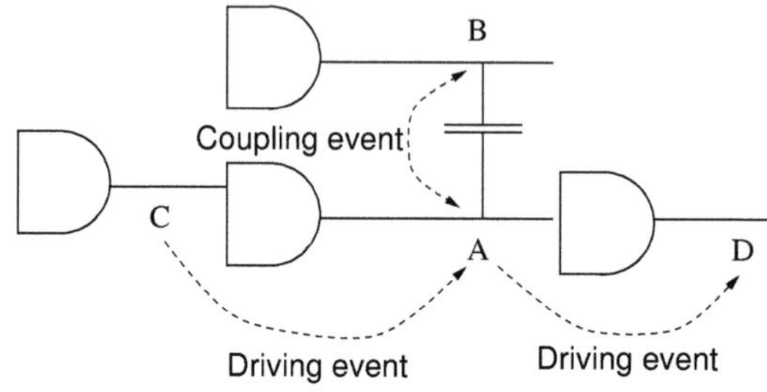

Figure 4.6.   Coupling/Driving Events

Updating

1 Schedule a coupling event for each node.

2 Pop an earliest event until the queue is empty according to the current status of the circuit

   (a) If it is a coupling event from node B to node A, as shown in Figure 4.6, compute the superposition coupling waveform to get the coupled delay of node A. Schedule the next stage driving events for node D.

   (b) If it is a driving event from node C to node A, update the propagation delay of the current node A, schedule a driving event from node A to node D, force a coupling event on node A to recompute the coupling effect, and check if node A attacks the adjacent nodes again. Schedule the coupling events for attacked adjacent nodes – for example, a coupling event from node A to node B.

Our algorithm keeps track of all old delay values. If a coupling or a driving condition does not change, it is not necessary to recompute or schedule an event.

4.2 Convergence of Our Algorithm

Intuitively, our algorithm tries to maintain a consistent system that each node has its propagation min and propagation max delays, as defined by Eq. 4.1 and 4.2, and the coupled delay conforms to the worst case alignment of its aggressor waveforms, as described in Section 4.4. Our algorithm corrects the local inconsistency of delays, and issues the related delay perturbation event to the next stage delays or the adjacent coupling delays.

THEOREM 4.2 *Given an accuracy requirement, the algorithm described above converges to a consistent value for each delay in a circuit using a finite number of steps.*

(Proof:) If there is a coupled delay inconsistency or a driving delay inconsistency on a node, our algorithm, as described in Section 4.4 and Eq. (1) or (2), recomputes it according to the coupling nodes or the incoming driving delays. This algorithm updates its coupled or propagation delay, and issues events for updating related coupling nodes and the next stage nodes. This maintains local consistency. Note that we assume each isolated sub-circuit group has at least one input to initiate the event-driven process. Initially, we assume the propagation delay for each primary input is fixed.

We now prove the algorithm's convergence. Supposing there is no coupling, we can compute the propagation delays in a topological order in one single pass. However, due to crosstalk coupling, a victim node might have coupling from its transitive fanouts whose switching windows cannot be finalized at the time when we calculate the propagation delay of the victim. Suppose one of the aggressor nodes j is a transitive fanout of a victim node i, and their switching windows overlap each other. Figure 4.7 shows these waveforms. We will prove these converge to a single point. As aggressor j moves from left to right, we can plot $t_i^{cpl,min}$ as a function of $t_j^{ppg,min}$ as shown in Figure 4.8, where T is a shorthand for $t_i^{ppg,min}$, h_j is the peak noise from aggressor j, $d_{i,j}^*$ is a transitive delay from node i to node j, and the propagation delay of node j

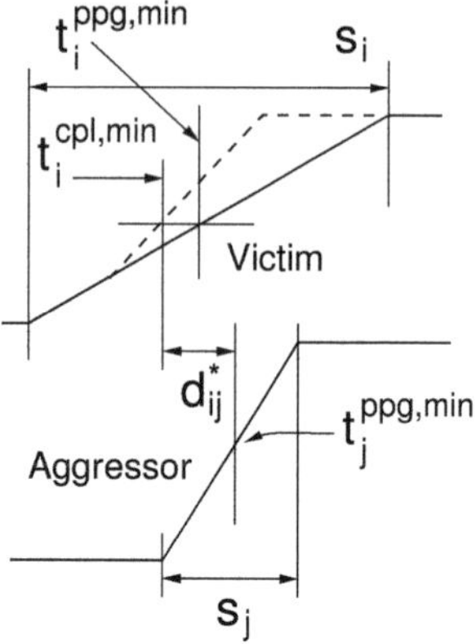

Figure 4.7.   Transitive Fanout as an Aggressor

is equal to:

$$t_j^{ppg,min} = t_i^{cpl,min} + d_{i,j}^* \qquad (4.3)$$

This function means that it can be sped up to the lower bound of $T - s_i h_j$ and it has an upper bound of T, which is not sped up. In addition, Eq. 4.3 has a lower bound when $d_{i,j}^*$ is equal to zero. This convergence process is shown in Figure 4.9. If at first $t_i^{cpl,min}$ is at point a, the value of $t_j^{ppg,min}$ can be obtained by Eq. 4.3. Then, the value of $t_i^{cpl,min}$ is obtained by the $t_i^{cpl,min}$ function shown in Figure 4.8, which is point b. It will continue this process to points c, d, and e, until it converges to the cross point z, which meets the accuracy requirement.

Moreover, the slope of the middle linear segment of $t_i^{cpl,min}$ function can be shown to be greater than 1, which means we can have only one cross point, since Eq. 4.3 is also linear. Thus, the iteration process must be able to improve towards convergence. With an accuracy requirement, we can reach this requirement in a finite number of steps. $\qquad\qquad \square$

Note that the iteration occurs when the aggressor's sensitive window overlaps the victim's switching threshold point, and the aggressor is one of the transitive fanouts of the victim. As the transi-

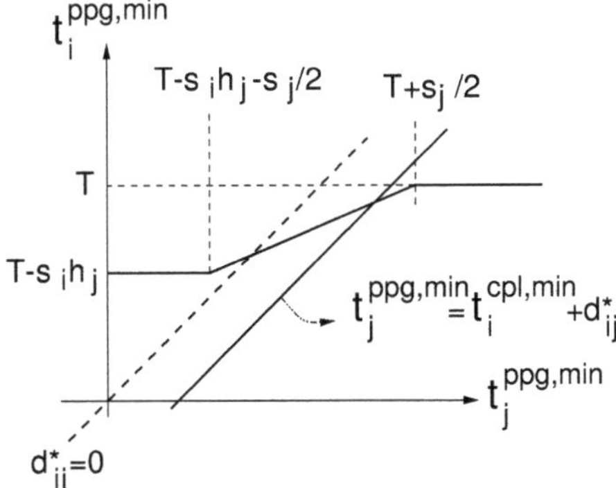

Figure 4.8. $t_i^{cpl,min}$ Function of $t_j^{ppg,min}$ and convergence

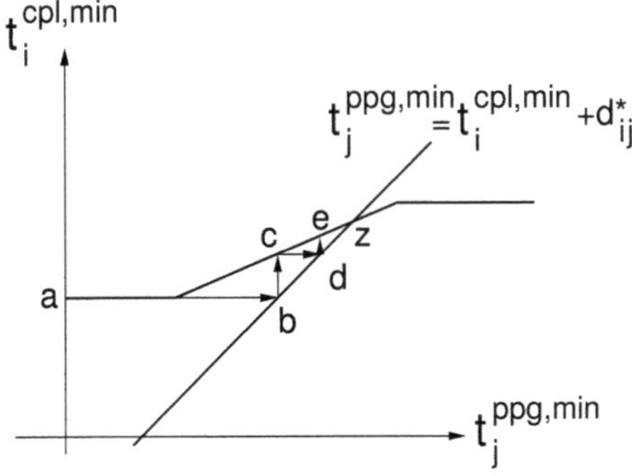

Figure 4.9. Convergence of $t_i^{cpl,min}$ and $t_j^{ppg,min}$

tive delay is shorter and the aggressor's slew is longer, it is likely to overlap and increase the computation time.

For the max delay, because of the conservative assumption that the aggressor can switch at any time point within the switching window to create the worst case coupling, the result of the event-driven algorithm always takes the worst case timing, which is conservative and no iteration is necessary. The $t_i^{cpl,max}$ function has two types, as shown in Figure 4.10.

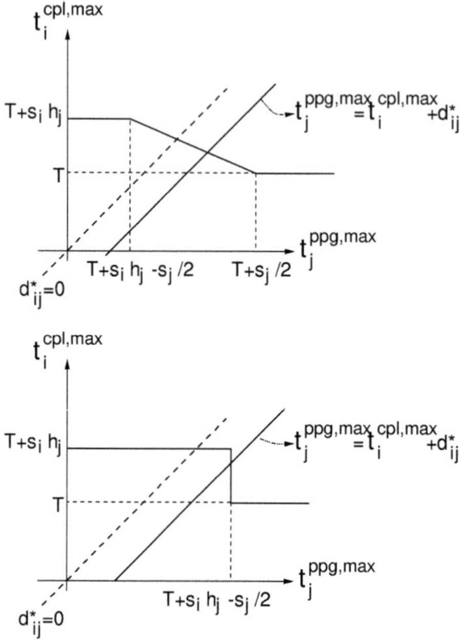

Figure 4.10. $t_i^{cpl,max}$ Function of $t_j^{ppg,max}$

4.3   Properties of Our Algorithm

It is interesting to note that our algorithm can reach the same result even if the initial values (propagation and coupled delay) are completely different. That is to say, our algorithm results in a very robust calculation. Different initial value settings only affect the number of events and calculation time. Typically if the initial values are closer to the final converged result, the computation time is reduced, because the algorithm converges in fewer steps.

4.4   Event Pruning

Any event causes a series of computations to update the whole system. However, coupling and driving events might not be necessary if they will not change any delay value of a circuit. Therefore, it is desirable to reduce the number of events issued to speed up the computation.

Coupling could be considered harmless if two signals switch with non-overlapping timing windows – that is, due to the temporal isolation, the two nodes that are physically coupled will result in no crosstalk noise effect. We can also compute the lower bound of min delay and upper bound of max delay by 0 or 3X coupling capacitance. It can be done before running the event-driven algorithm, and provides valuable information for pruning events.

When a node changes its propagation delay, our algorithm issues events based on the following facts:

1 If the coupling computation in some previous event still keeps the same condition for coupling, it does not need to schedule this event, since the same coupling condition results in the same amount of coupling noise.

2 When the coupling condition changes, a coupling event must be issued to update the corresponding coupling nodes, and the propagation delay of a next stage can change accordingly, so a driving event is issued.

The event-driven type of calculation makes the computation very robust and efficient with reduced re-computation.

4.5 Scheduling Technique

Scheduling is a key for the efficiency of convergence. We can reduce complexity by an order of magnitude with careful arrangement of events. We identify the following scheduling approaches:

Dynamic Event Time We schedule events based on the right hand bound of sensitive windows, defined in Section 4.2, and dynamically sort the events according to the current circuit status (delay value). Intuitively, it forms a sweep timing line across the circuit. If any event occurs earlier in the event time, our event algorithm schedules it first and continues iterating on its related events until it converges.

Static Event Time We schedule events based on the right hand bound of sensitive windows. No dynamic sorting of events is performed. When two events have the same time, we schedule events based on the topological order of the nodes.

Smart Global We maintain a flag to identify whether one node must be updated or not. At each pass, each node is examined and processed if necessary. The updated delay will propagate to its coupling nodes and next stages. The event pruning technique is also used to reduce the number of updates. If no update is needed throughout a pass, the computation has converged.

5.   Experimental Results

We demonstrate our algorithm on a 233MHz PC with 64M bytes of memory running the Linux operating system. We benchmark our algorithm on the ISCAS85 combinational circuits. For every circuit, each node is presumed to have four randomly chosen coupling nodes. The coupling noise between each pair of aggressors and victims, and the slew on each node are pre-characterized or estimated. We also vary these parameters with different scheduling approaches to test the efficiency of our algorithm.

The total run time for all of the ISCAS85 11 combinational circuits takes only 7.09 seconds. It is observed that 21.9% of the nodes, on average, are recomputed for coupling calculation, which means only 21.9% of the nodes have to be calculated twice for the coupling to obtain to the final delay value. Table 4.1 shows this result, where the first column is the name of circuit, the second column is the number of nodes, the third column is the total number of fanouts, which is equal to the number of driving edges, the fourth column is the number of coupling computations, the fifth column is the percentage of re-computation of coupling, and the last column is the run time.

Circuit	#Nodes	#Edges	#Coupling Comp.	Run Time
C17	11	12	23	0.005s
C432	196	336	424	0.083s
C499	243	408	538	0.099s
C880	443	729	1042	0.200s
C1355	587	1064	1361	0.276s
C1908	913	1498	2041	0.441s
C2670	1350	2076	3082	0.638s
C3540	1719	2939	3924	0.880s
C5315	2485	4386	5477	1.225s
C6288	2448	4800	7281	1.350s
C7552	3718	6144	9220	1.896s

Table 4.1.   Result for ISCAS85 Combinational Circuits

W factor	#Coupling Comp.	#ReComp.%	Run Time
1.0	40528	43.6%	7.55s
0.8	37192	31.8%	7.31s
0.6	34467	22.1%	7.17s
0.5	34413	21.9%	7.09s
0.4	35080	24.3%	7.17s
0.2	37121	31.5%	7.21s
0.0	41828	48.2%	7.39s

Table 4.2.   Initial values affects the number of coupling computations

With different initial values, the number of coupling computations can have a 22% difference, as shown in Table 4.2. The first column shows "W": factor, which is the factor of how close the initial value is to the worst case value: 0.0 represents using the nominal delay value, and 1.0 represents using the worst case value for initial values. The second column is the total number of coupling computations for all 11 combinational circuits from ISCAS85. The third column is the percentage of re-computations, and the last column is the run time.

Scheduling factor	#Coupling	Run
Method	Comp.	Time
Smart Global	240292	96.8s
Dynamic Event Time	308048	113.2s
Static Event Time	286718	108.2s

Table 4.3.   Performance for Different Scheduling Approaches

Ckt	#Nodes	#Edges	#Cpl. Edges	#Cpl. Comp.	Run Time
c17	22	20	48	184	0.01s
c432	709	794	3049	4769	1.31s
c499	1239	1372	9757	5319	1.59s
c880	2382	2568	21026	12002	3.59s
c1355	1655	1796	13110	7204	2.15s
c1908	1353	1473	11113	6177	1.81s
c2670	2760	2870	30223	12719	5.58s
c3540	4650	5077	52341	22376	10.10s
c5315	6093	6647	63232	35871	12.74s
c6288	21153	24375	245889	109308	45.21s
c7552	8630	9445	83962	43846	20.54s

Table 4.4.   Results for $0.25\mu m$ process implementation of ISCAS85 combinational circuits

We also implement the ISCAS combinational circuits in a $0.25\mu m$ technology. The result is as shown in Table 4.4, where the first column shows the circuit name, the second column is the number of nodes in the circuit, the third column is the number of propagation edges, the fourth column is the number of the coupling edges, the fifth column is the number of coupling computations, and the last column is the run time. In Table 4.3, we compare different scheduling approaches in terms of total run time on all of these circuits. The Smart Global scheduling approach is the winner among all of the scheduling approaches.

6. Review of Conservativism

If the initial switching windows are infinity, according to Section 3.3.4, we can get upper bounds or conservative results through out the iterations and converge into a final solution. The scheduling methods proposed in this chapter can speed up the iterations. If the initial switching windows begin from the nominal ones, we cannot get any valid upper bound until the iteration reach the fixed point described in Chapter 3.

7. Conclusion

Using a flexible and practical waveform model, We propose a robust and efficient algorithm to compute the coupling delay effect on static timing analysis. This approach can be directly implemented in a very practical industrial tool for advanced static timing analysis targeting very deep submicron designs.

Chapter 5

REFINEMENT OF SWITCHING WINDOWS

Chapter 4 introduced a technique to compute switching windows to improve the computation time of crosstalk calculation. In this chapter, we introduce a method to further reduce pessimism of crosstalk analysis based on time slots. For crosstalk noise calculations, computing switching windows of a net helps identify noise sources accurately. Traditional approaches use a single continuous switching window for a net. Under this model, it is assumed that signal switching happens at any time within the window. Although conservative and sound, this model can result in too much pessimism because the exact timing of signal switching is determined by a path delay up to the net (i.e. the underlying circuit structure does not always allow signal switching at arbitrary times within the continuous switching window). To address this inherent inaccuracy of the continuous switching window, we propose a refinement of the traditional approaches, such that signal switching is characterized by a set of discontinuous switching windows instead of a single continuous window. Each continuous switching window is divided into multiple windows, called time slots, and the signal switching activity of each slot is analyzed separately to calculate the maximum noise more accurately. By controlling the size of a time slot, we can trade off accuracy and runtime, which makes this approach scalable for large designs. We have confirmed by experiments on industrial circuits that up to 90% of the noise violations detected by the traditional approach can be unreal.

1.   Introduction

Recall that crosstalk noise affects timing by either decreasing or increasing the delay. If the adjacent nets are quiet, there is no crosstalk noise. Therefore, it is important to identify the switching window (or timing windows), so that if there is no overlap of switching windows between two coupling nodes, we can immediately conclude that there is no timing variation, thereby reducing the analysis pessimism.

Typically, the switching windows considered in the literature [Sap99, FFP$^+$00, TCE00, ARP00, CKK00b, XCMS00, ZSN01, CKTK02] are continuous (see Figure 5.1). They are a timing interval from the earliest arrival time to the latest arrival time of a net. However, because the number of possible timing paths to a net is bounded by the number of topological circuit paths, the arrival times are typically not continuous inside a switching window (see Figure 5.2); instead, they are discrete arrival times. Consider Figures 5.1 and 5.2. Figure 5.2 captures switching activity more accurately by discontinuous windows, while Figure 5.1 is an approximation of Figure 5.2 by a continuous window. Suppose net A and net B are aggressors to be aligned for the maximum noise. As shown in Figure 5.2, there is no switching window overlap. However, if Figure 5.2 is approximated by Figure 5.1; the two windows have overlap, resulting in a false alignment and a pessimistic noise estimation. The goal of this chapter is to take advantage of discontinuous switching windows to calculate crosstalk noise more accurately.

Using a fixed delay model, the maximum number of discrete arrival times at a net is equal to the number of topological paths to the net. To simplify the analysis, we ignore functional dependency in the following discussion throughout this chapter. To avoid han-

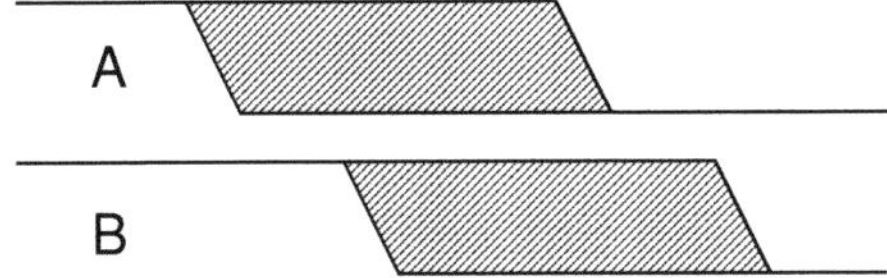

Figure 5.1. Continuous Switching Windows

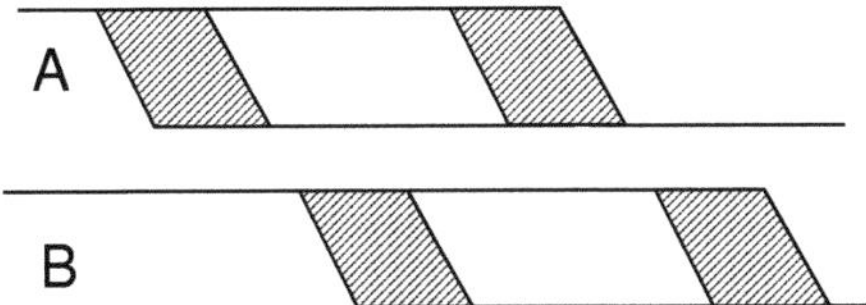

Figure 5.2. Discontinuous Switching Windows

dling a potentially exponential number of discrete arrival times, we use a time slot approach, where a continuous switching window is refined as a set of time slots of the same size. Thus, the size of a time slot is an effective scaling factor to trade off analysis accuracy versus speed and capacity. As the analysis resolution gets finer, more maximum noise can be precisely justified.

The rest of this chapter is organized as follows. In Section 5.2, we introduce the formulation and algorithmic aspects of our approach, and discuss the theory behind it. In Section 5.3, we address the resolution issues. In Section 5.4, we show the experimental results. In Section 5.5, we investigate a refinement of the proposed method, where slew effects on the maximum noise are modeled more accurately.

2. Formulation and Algorithm

Recall that **victim** is a net that suffers from a noise effect, and an **aggressor** is a net that contributes noise. Their roles can change depending on the context. Using the notations of Chapter 3, The latest and the earliest arrival times of net i can be written as:

$$x_i = \max_{k \in D_i}(x_k + d_{ki} + \delta_{ki}) \tag{5.1}$$

and:

$$y_i = \min_{k \in D_i} (y_k + d_{ki} + \delta_{ki}), \tag{5.2}$$

respectively, as shown in Figure 5.3. The traditional continuous

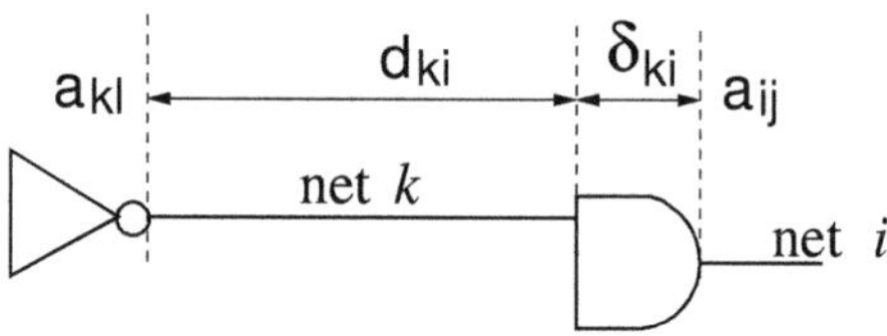

Figure 5.3. Gate delay and interconnect delay variables

switching window model defines the switching window from time y_i to time x_i.

Let s be the size of a time slot, and $a_{ij} \in \{0, 1\}$ be a Boolean variable defined as follows:

$$\begin{cases} a_{ij} = 1, & \text{if net } i \text{ has any arrival time } t \\ & \quad \text{such that } js \le t < (j+1)s. \\ a_{ij} = 0, & \text{otherwise.} \end{cases}$$

a_{ij} is used to record if net i has any signal arriving at time slot j. For multiple arrival times of a net i, the corresponding a_{ij}'s are non-zero. It can be recursively computed as:

$$a_{ij} = \bigvee_{k \in D_i} a_{kl}, \tag{5.3}$$

where $\bigvee$ is a logical "OR" operator and $l = j - \lfloor \frac{d_{ki} + \delta_{ki}}{s} \rfloor$. This is to say a_{ij} is non-zero if any of preceding stages has an arrival time within appropriate time slots. Strictly speaking, we have to check two time slots:

$$\lfloor j - \frac{d_{ki} + \delta_{ki}}{s} \rfloor \le l \le \lfloor j + 1 - \frac{d_{ki} + \delta_{ki}}{s} \rfloor.$$

If the gate delay δ_{ki} is considered within a range $[\delta_{ki}^{\min}, \delta_{ki}^{\max}]$, the time slots to check are:

$$\lfloor j - \frac{d_{ki} + \delta_{ki}^{\max}}{s} \rfloor \le l \le \lfloor j + 1 - \frac{d_{ki} + \delta_{ki}^{\min}}{s} \rfloor.$$

The nets are visited in the same order as in the continuous switching window calculation or longest path calculation: a depth-first traversal of the direct acyclic graph (DAG). Assuming that aggressors in the same time slot can align to create the maximum noise on the victim net i, we can then calculate the maximum noise at net i as:

$$\max_{0 \le j \le \lfloor \frac{L}{s} \rfloor} \sum_{k \in A_i} \phi_{ki} a_{kj}, \qquad (5.4)$$

where L is the longest path delay or the maximum arrival time of the circuit, and $\phi_{ki} \in R$ is the noise effect from the aggressor net k to the victim net i. Although the slew rate of net k might not be the same for all the time slots, ϕ_{ki} is calculated using the fastest slew available on net k to ensure a pessimistic analysis. We can further reduce the search range down to the interval $[\lfloor \frac{y_{A_i}^{\min}}{s} \rfloor, \lfloor \frac{x_{A_i}^{\max}}{s} \rfloor]$, where $y_{A_i}^{\min}$ is the minimum arrival time of all the aggressors of net i, and $x_{A_i}^{\max}$ is the maximum arrival time of all the aggressors of net i — that is, Eq. 5.4 becomes:

$$\max_{\lfloor \frac{y_{A_i}^{\min}}{s} \rfloor \le j \le \lfloor \frac{x_{A_i}^{\max}}{s} \rfloor} \sum_{k \in A_i} \phi_{ki} a_{kj}.$$

2.1 Arrival Time Uncertainty in Interconnect

Due to a signal transition through an interconnect, arrival times can be quite different between the driver and the receivers of a net. This interconnect delay can be captured as uncertainty of the arrival time for the net. Thus, the switching window of a net must be spread out to cover the uncertainty in interconnect signal propagation. For example, suppose the arrival time is 1000ps at the driver of a net, if the interconnect can take up to the maximum of 200ps to propagate to a receiver of the net, we have to mark the arrival time slots from 1000ps to 1200ns in the switching window of the net. During this time interval, the victims of this net might have crosstalk effects.

Let $d_k^{\max}$ be the maximum interconnect delay on net k. We must spread out the interconnect as:

$$\begin{cases} a_{im}^+ = 1, & \text{if } a_{ij} = 1 \text{ and } j \le m \le j + \lfloor \frac{d_k^{\max}}{s} \rfloor, \\ a_{im}^+ = 0, & \text{otherwise}, \end{cases}$$

where a_{im}^+ is the modified arrival time slot m for net i. The maximum noise is thus revised as:

$$\max_{0 \le j \le \lfloor \frac{L}{s} \rfloor} \sum_{k \in A_i} \phi_{ki} a_{kj}^+. \tag{5.5}$$

Eq. 5.1 needs to be revised as:

$$x_i^+ = d_i^{\max} + \max_{k \in D_i}(x_k + d_{ki} + \delta_{ki}).$$

Note that x_k still remains the same, and does not affect the earliest arrival time.

2.2   Switching Window Density

Compared with the traditional approach using continuous switching windows, the time slot approach can help reduce analysis pessimism. The effectiveness strongly depends on the switching windows' density, which can be defined as the ratio of the number of non-zero time slots to the total number of time slots in a switching window. The traditional continuous switching windows have a density of 1 by definition. This density measure is a very good metric of how effective the time slot approach reduces the pessimism of noise analysis. If the density is close to 0, the switching window tends to be very sparse, and the time slot approach can cut down most of the pessimism in the maximum crosstalk noise calculation.

2.3   Input Timing Uncertainty

Typically, the arrival time of an input pin of a chip is given not as a constant but a bounded timing range. It could represent the timing uncertainty due to the process, voltage, or temperature variation.

Let I be the index set of input nets. For $i \in I$, we have

$$\begin{cases} a_{ij} = 1, & \text{if } a_{ij} = 1 \text{ and } \lfloor \frac{y_i}{s} \rfloor \le j \le \lfloor \frac{x_i}{s} \rfloor, \\ a_{ij} = 0, & \text{otherwise,} \end{cases}$$

where y_i and x_i are the earliest arrival time and the latest arrival time, respectively. The larger the timing uncertainty, the denser the switching windows. Therefore, reducing the timing uncertainty at the inputs can increase the effectiveness of our approach.

2.4   Complexity

Given the N nets and the total M fanouts of nets in a circuit (similar to a direct acyclic graph with N vertices and M edges), the complexity of calculating the arrival time slots is $O(N + M\lfloor \frac{L}{s} \rfloor)$ by Eq. 5.3. The dominant operation is actually the maximum noise calculation by Eq. 5.5, which has the complexity $O(NP\lfloor \frac{L}{s} \rfloor)$, where P is the maximum number of aggressors of a net, or equivalently the maximum cardinality of A_i, where $i = 1 \ldots N$.

2.5   Implementation Consideration

Since a_{ij}'s are Boolean variables, we can compact them into a 32-bit integer. a_{ij} can be easily represented by a bit map. The time slot ranges from $\lfloor \frac{y_i}{s} \rfloor$ to $\lfloor \frac{x_i^+}{s} \rfloor$ for a net i. This approach has very efficient memory usage with a slight speed penalty.

3.   Resolution and Truncation Errors

The size of the time slot is an important factor for the analysis accuracy. We test different slot sizes on a small circuit with 8828 nets and 7956 cell instances using $0.25\mu m$ technology, and check the maximum noise peak over 20% of power voltage. Table 5.1 shows the slot size effect, where the first column shows the slot size, the second column shows the number of noise violations and the third column shows the switching window density. The second

Slot size	#Violation	S.W. Density
Continuous	352	1.000
400ps	204	0.891
200ps	162	0.842
100ps	131	0.789
90ps	134	0.779
80ps	132	0.755
70ps	130	0.720
50ps	123	0.720
1ps	94	0.430

Table 5.1.   Slot size effect on the number of noise violations.

"Continuous" row is the traditional continuous switching window method. In general, a smaller slot size can result in fewer noise violations. However, notice that the slot size 90ps generates more violations than the slot size 100ps.

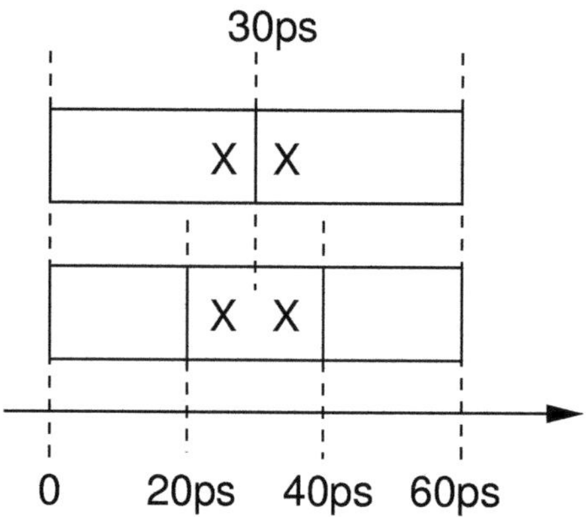

Figure 5.4.   Finer slot gets more noise violations

Consider the case in Figure 5.4, where the time slot sizes are 20ps and 30ps, respectively. Suppose the two aggressors' arrival times are 25ps and 35ps, respectively. If the time slot size 20ps is used, we can align these two aggressors by assumption. If the time slot size 30ps is used, these two aggressors cannot align to create the maximum noise. Therefore, the finer time slot does not always

Circuit	#Nets	#Inst.	Slot size	S.W. Density	#Vio.	Run time
			Cont.	1.000	51	16.8s
designA	19.0K	17.5K	1000ps	0.676	41	16.7s
			100ps	0.596	29	17.2s
			10ps	0.470	29	25.4s
			Cont.	1.000	2652	49.8s
designS	25.0K	32.2K	1000ps	0.388	1553	46.1s
			100ps	0.257	1526	46.3s
			10ps	0.058	1507	47.4s
			Cont.	1.000	2815	54.8s
designC	55.8K	50.0K	1000ps	0.785	2210	53.6s
			100ps	0.647	1540	51.7s
			10ps	0.384	1519	54.7s
			Cont.	1.000	7330	176.0s
designB	81.8K	73.0K	1000ps	0.675	4738	176.0s
			100ps	0.596	2951	170.0s
			10ps	0.415	2430	245.0s
			Cont.	1.000	14091	295.0s
designV	273.4K	234.3K	1000ps	0.816	7094	259.0s
			100ps	0.614	3942	237.0s
			10ps	0.596	2831	266.0s
			Cont.	1.000	31782	2118.0s
designW	527.1K	444.7K	1000ps	0.722	10515	1927.0s
			100ps	0.512	3164	1832.0s
			10ps	0.270	3105	1993.0s

Table 5.2. Comparison of continuous switching window and time slot approach on the number of noise violations.

imply fewer noise violations. However, if the size of a coarser time slot is a multiple of a finer time slot's size, this can be avoided. So, as long as the time slot size is continuously divided evenly, as in Table 5.2 (1000ps $\rightarrow$ 100ps $\rightarrow$ 10ps), this problem disappears.

4.   Experimental Results

We conducted experiments on several industrial circuits of significant size. Table 5.2 shows the results. The first column is the circuit name. The second column is the number of nets in that circuit. The third column is the number of cell instances. The fourth column is the time slot size, where "Cont." represents the traditional continuous switching window approach. The fifth column is the switching window density. The sixth column shows the number of noise violations, where 40% of VDD is the threshold. The last column shows the run time on a Linux machine with 1.26GHz CPU.

From Table 5.2, the number of noise violations can be reduced dramatically by 90% for designW and 43% for designA. The run time penalty was a slight increase. In fact, the continuous switching window approach could take a longer run time due to the processing of an excess of noise violations. A finer time slot can reduce the number of the noise violations. However, the amount of reduction tends to decrease significantly after some finer time slots.

5.   Consideration of Slew Rates

The approach proposed above uses the fastest slew at net i in the computation of ϕ_{ki}. Basically, we then assume that this maximum noise effect ϕ_{ki} is achievable in any time slot in the entire switching window. This is a conservative assumption, but a slew at net i could be much larger than the fastest slew in some time slots, resulting in pessimism. We can incorporate this effect by propagating, and maintaining the fastest slew of each time slot, and computing ϕ_{ki} of each slot based on the slew. Let $\alpha_{ij} \in R$ be the minimum or fastest slew in time slot j of net i. The propagated slew model is thus written as:

$$\alpha_{ij} = \min_{k \in D_i} \alpha_{kl}$$

where $l = j - \lfloor \frac{\delta_{ik} + d_k}{s} \rfloor$. The maximum noise is calculated as:

$$\max_{0 \le j \le \lfloor \frac{L}{s} \rfloor} \sum_{k \in A_i} \Phi_{ki}(\alpha_{kj})$$

where $\Phi_{k,i} : R \to R$ is a noise peak function of slew rate, representing the noise effect from the aggressor net k to the victim net

i. Note that this approach has a heavy speed and space penalty because the bit pattern representation for a_{ij} cannot be used any longer and we must record slew information α_{ij} in each time slot. The number of computations of interconnect and gate delay is proportional to $O(NP\lfloor \frac{L}{s} \rfloor)$. This is not considered practical for multi million-gate designs.

6. Property of Time Slots and Conservativism

The approach in this chapter is actually discretization of a switching window. If this approach is applied to switching window calculation as we did in Chapter 3 and 4, clearly, it can create multiple solutions and even oscillate during the switching window calculation, because we use a Boolean variable to record if a signal arrives at a time slot. A possible improvement is to use multiple continuous switching windows for a net. The overlapping switching windows should be merged to reduce the complexity. However, as the number of possible timing paths may be exponential, this approach implies a costly run time penalty.

To make this time slot approach conservative, we need to make sure every timing uncertainty source (e.g. Section 5.2, 5.2.1 and 5.2.3) is considered in the time slot marking procedure. For example, if a net has a transition time which is larger than the time slot size, we need to mark every time slot that has overlapping with the transition. Another option is to store a fractional number instead of a Boolean variable in each time slot to record this partial transition information.

7. Conclusion

Switching windows can be refined by time slots to improve accuracy. Our experiments show that up to 90% of potential noise violations detected by continuous switching windows can be excluded by this approach. Moreover, the size of a time slot can be controlled to trade off accuracy for speed and capacity, which makes this algorithm scalable for industrial-sized circuits.

Chapter 6

FUNCTIONAL CROSSTALK ANALYSIS

We now consider the use of functional information in crosstalk analysis. The goal of this chapter is to develop an algorithm and noise analysis flow that provide an accurate and conservative approach to functional crosstalk analysis. In particular, this chapter proposes an approach to identifying a pair of vectors that exercises the maximum crosstalk noise.

1. Introduction

The current approaches to interconnect crosstalk analysis are based on identifying the spatial relationship between two coupling signals, and then adding a *static analysis* of the temporal relationship [She98a][Kir97]. The use of static timing information in this methodology is similar to the static timing analysis without false-path elimination, so it might lead to an overly pessimistic estimation of the actual noise in the circuit. In the case of false noise analysis, the practical impact is wires' re-routing or signal drivers' modification. On the other hand, a greater drawback of static analysis approach is that it might fail to correctly analyze signal glitches, which can also be responsible for erroneous switching in the circuit.

One key to improving the accuracy in static timing analysis is to add functional information and thereby compute the "true delay" of the circuit. Similarly, we believe the key to improving noise analysis is also to add functional information to the temporal. By analogy, we call this computing the "true noise" of the circuit.

In this chapter, we will present a method for searching for the vector pair which maximizes the crosstalk noise on a given net for a combinational sub-circuit. This search uses the timing information for the relevant signals together with the functional information of the gates computing the signals. These two together form a tighter upper bound, and the input vectors that exercise the maximum crosstalk noise can be identified.

In Section 2, we review some useful methods and related work for crosstalk analysis. In Section 3, we introduce some techniques to prune this large vector search space. In addition, we will discuss several models with respect to their complexity, efficiency, and accuracy for computing the noise bound. In Section 3, we explain the vector search algorithm for the maximum crosstalk noise is explained. Section 4 shows the experimental results, and in Section 5, some areas for improvement are discussed.

2.   Approaches and Related Work

The most straightforward approach to finding a vector pair that maximizes the noise on a given net is to exhaustively simulate all input vector pairs. This is rarely computationally feasible. Therefore, an accurate and computationally efficient method needs to find the vector pair that stimulates the maximum noise. Instead of exhaustive logic simulation, we formulate this problem as a Boolean Constrained Optimization Problem (BCOP) to solve it exactly. Moreover, our approach can be extended to a general vector-search scheme, which can be used to search for the vectors causing the maximum IR drop or the maximum power consumption [DKW92].

We will begin by reviewing prior work that has some elements in common with the the problem of maximum noise analysis. In [RIZK94], the authors propose a multiple-value logic to generate an input vector to test cross coupling faults. However, there is no timing involved, and thus it is unable to find the real noise bound – only a pair of vectors for circuit testing.

In [DKW92], the authors provide an approach to finding a vector pair which maximizes power dissipation. This problem shares a couple of elements in common with our own: function and timing must be integrated, and failure to analyze glitches will lead to a non-conservative procedure. Unfortunately, the search approach shown in [DKW92] is somewhat primitive and is unable to scale to the size of problem we want to consider.

The use of Timed Boolean Functions (TBF) [LB94] has also been proposed for computation of the vector pair that causes the maximum number of transitions. This is related to our problem, but it is not equivalent and no strategy for noise analysis is reported. The underlying computational mechanism is based on Binary Decision Diagrams (BDDs) [Bry86]. Also, basing his approach on BDDs is [Kir97]. This work does focus on noise analysis and does a nice job of describing the relationship between spatial, temporal, and functional elements of noise analysis; however, in [Kir97] results are cited on only very modestly sized circuits ($<$ 1000 gates). BDD-based techniques are necessarily limited because many circuits do not have compact BDD representations.

Other recent work has used Timed Automata to model coupling delays [ea97]; however, these approaches are even more computationally expensive than the BDD-based approaches.

In conclusion, although a variety of work has been performed that is relevant to this problem, there is still a need for an accurate "true noise" analysis approach that is computationally efficient. This is the goal of this chapter.

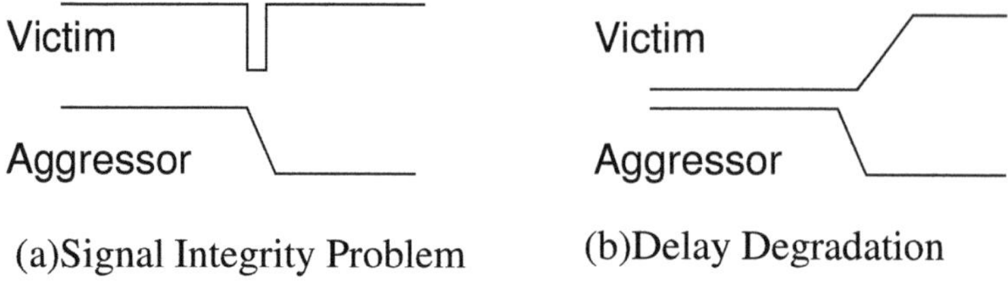

Figure 6.1.   Signal Integrity and Delay Degradation

3.   Vector Pair Searching Algorithm
3.1   Overview

Our proposed approach is based on a similar method of functional timing verification [DKM93, DKMW94, Sas93]. We need to set up timed Boolean variables for each node at some time points and conjunct the gate characteristic functions, which represent all the possible logic combinations of a gate. According to the allowable noise level to assign the weights of some Boolean variables, the vector pair search problem can be formulated as a BCOP, in which we find some assignment to satisfy the constraints and maximize an objective function in terms of these Boolean variables. In the following, we describe each step in detail.

3.2   BCOP: Boolean Constrained Optimization Problem

Given an objective function,

$$h(x_1, x_2, \ldots, x_n) = c_1 x_1 + c_2 x_2 + \cdots + c_n x_n,$$

where $x_1, x_2, \ldots, x_n \in B = \{0, 1\}$ and $c_1, c_2, \ldots, c_n \in R$, we want to maximize $h(x_1, x_2, \ldots, x_n)$ such that it satisfies the constraint:

$$f(x_1, x_2, \ldots, x_n, \overline{x_1}, \overline{x_2}, \ldots, \overline{x_n}) = \prod_i \sum_j x_{ij} \equiv 1,$$

where $x_{ij} \in \{x_1, x_2, \ldots, x_n, \overline{x_1}, \overline{x_2}, \ldots, \overline{x_n}\}$. The constraint can be written in Conjunction Normal Form(CNF). This problem is similar to the BCP.

For weighting a switching on the aggressor nets, the cost function might not be linear. Moreover, if timing is considered, some weighted terms must be represented as a complex conjunction of some circuit variables. In such cases, we implement a branch-and-bound algorithm to address this special objective function evaluation.

3.3 Constructing Circuit via SAT

To represent a combinational circuit, we use the characteristic function of each gate and set up a variable for each node. Conjuncting these characteristic functions together, we obtain a characteristic function to represent the whole circuit [Lar92][SBSV96]. These characteristic functions are represented by CNF clauses. A circuit is functionally consistent if and only if the CNF clause can be satisfied. Namely, we can find a valid assignment for each variable, which means there is a valid logic combination of the inputs, the outputs and the internal nodes of the circuit.

For example, in Figure 6.2, this AND circuit can be characterized by

$$f(X, A, B) = (X + \overline{A} + \overline{B})(\overline{X} + A)(\overline{X} + B). \qquad (6.1)$$

f is equal to one if and only if X, A, and B are a valid variable

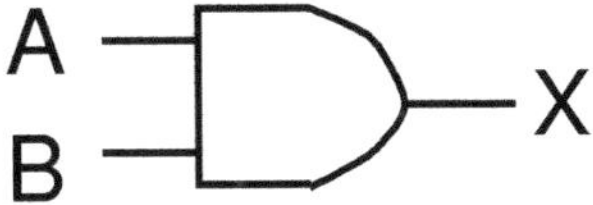

Figure 6.2. Characteristic Function of an AND Gate

assignment for AND gate operation. Consider the circuit in Figure 6.3. We can conjunct the characteristic functions of all gates together to obtain the characteristic function of this whole circuit.

$$\begin{aligned}
f(Z, X, Y, A, B, C) &= (X + \overline{A} + \overline{B})(\overline{X} + A)(\overline{X} + B) \\
&\cdot (Y + C)(\overline{Y} + \overline{C}) \\
&\cdot (\overline{Z} + X + Y)(Z + \overline{X})(Z + \overline{Y}).
\end{aligned}$$

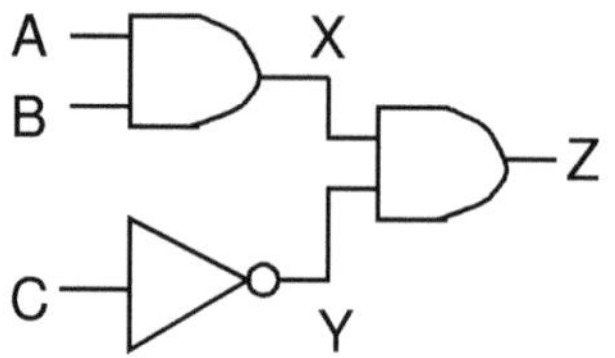

Figure 6.3. Conjunction of Characteristic Functions

Thus, the stable state of this circuit must satisfy the constraint above (f equals 1).

3.4 Maximum Noise under the Zero-Delay Model

In the zero-delay model, all the gates and the interconnect are assumed to have zero delay. Therefore, the maximum noise occurs when all the aggressor nets make transitions in the same direction. We can then investigate the correlation between signals to find the maximum number of the opposite transitions, which maximizes the crosstalk noise under the zero-delay model.

For a single victim net, given the coupling capacitance between each aggressor net and the victim net, we can obtain the maximum noise by:

1 Setting up two variables for each node: one variable denotes the value of time 0, and the other is for time ∞. We denote a variable for node X as X_0 and X_∞, respectively for time 0 and time ∞.

2 Build the characteristic function into the CNF clause for two sets of variables. For the circuit in Figure 6.3, we have:

$$
\begin{aligned}
1 \equiv\ & (X_0 + \overline{A}_0 + \overline{B}_0)(\overline{X}_0 + A_0)(\overline{X}_0 + B_0) \\
& \cdot\ (Y_0 + C_0)(\overline{Y}_0 + \overline{C}_0) \\
& \cdot\ (\overline{Z}_0 + X_0 + Y_0)(Z_0 + \overline{X}_0)(Z_0 + \overline{Y}_0) \\
& \cdot\ (X_\infty + \overline{A}_\infty + \overline{B}_\infty)(\overline{X}_\infty + A_\infty)(\overline{X}_\infty + B_\infty) \\
& \cdot\ (Y_\infty + C_\infty)(\overline{Y}_\infty + \overline{C}_\infty) \\
& \cdot\ (\overline{Z}_\infty + X_\infty + Y_\infty)(Z_\infty + \overline{X}_\infty)(Z_\infty + \overline{Y}_\infty).
\end{aligned}
$$

3 Find the maximum (or minimum) weighted sum of each aggressor's variable, where the weighting is proportional to the coupling capacitance or the delta voltage on the victim caused by the aggressor's switching. The phases are assigned according to the switching direction on the aggressor nets. Suppose we have node B and Y as aggressors of node Z in Figure 6.3. The objective function to maximize or minimize is:

$$V_{B,Z}\overline{B}_0 B_\infty + V_{Y,Z}\overline{Y}_0 Y_\infty - V_{B,Z} B_0 \overline{B}_\infty - V_{Y,Z} Y_0 \overline{Y}_\infty,$$

where $V_{B,Z}$ is the delta voltage at node Z caused by node B switching, and $V_{Y,Z}$ is the delta voltage at node Z caused by node Y switching.

4 Set up the initial condition that makes the victim net static – that is, the victim net stays on the same logic state even after the second vector is applied. For the circuit in Figure 6.3, we have:

$$1 \equiv Z_0 Z_\infty + \overline{Z}_0 \overline{Z}_\infty.$$

3.5 Fixed Delay Circuit Construction via SAT

The zero-delay model, in general, does not give an accurate noise bound, and it often gives almost the same scale of the noise bound as the static noise analysis. Thus, timing information is important to be considered in this vector search procedure to obtain a tighter maximum noise bound.

Timing information can be included by introducing timed Boolean variables, with which we can represent the logic values of a node at different time points.

Also, it is assumed that there are 2 input vectors to apply to the primary inputs of a circuit: one vector at time minus infinity, and the other at time zero. The former vector drives each node of the circuit into a known state no matter how long the circuit delay is, and the latter exercises the maximum crosstalk noise. This scheme can be easily generalized for multiple vectors or extended to the other applications.

3.5.1 Using Timed Boolean Variables

The timed Boolean variable, used to represent a logic state for a node at some time point, is a mapping $v : R \rightarrow \{0, 1\}$. The negative timed Boolean variable denoted by $\overline{v}$ is the complemented variable of v.

Suppose the circuit in Figure 6.4, where $Z(t = 4)$ is evaluated.

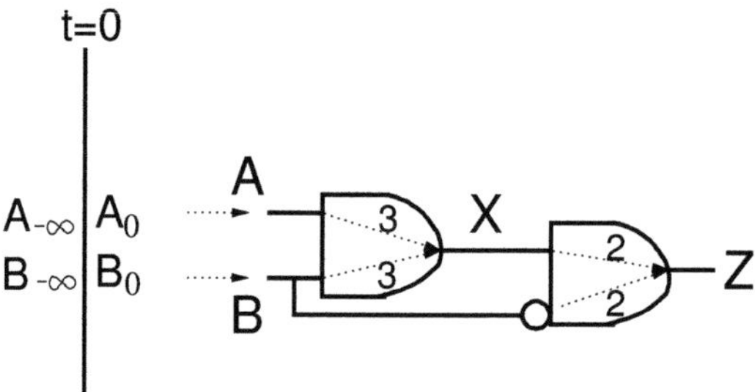

Figure 6.4.   Timed Boolean Variable Example

$Z(t = 4)$ is abbreviated as Z_4. Since the delays between X to Z and B to Z are 2, we can represent the gate on the right by the characteristic function:

$$(Z_4 + \overline{X}_2 + B_2)(\overline{Z}_4 + X_2)(\overline{Z}_4 + \overline{B}_2)$$

Since B is a primary input, only two values are applied. B_2 is equal to B_0. We rewrite the above characteristic function as:

$$(Z_4 + \overline{X}_2 + B_0)(\overline{Z}_4 + X_2)(\overline{Z}_4 + \overline{B}_0). \qquad (6.2)$$

Similarly,

$$(X_2 + \overline{A}_{-1} + \overline{B}_{-1})$$
$$(\overline{X}_2 + A_{-1})(\overline{X}_2 + B_{-1})$$

A and B are primary inputs, $A_{-1} = A_{-\infty}$ and $B_{-1} = B_{-\infty}$. Therefore,

$$(X_2 + \overline{A}_{-\infty} + \overline{B}_{-\infty})$$

$$(\overline{X}_2 + A_{-\infty})(\overline{X}_2 + B_{-\infty}). \tag{6.3}$$

The whole circuit characteristic function for $Z(t = 4)$ becomes

$$(Z_4 + \overline{X}_2 + B_0)$$
$$(\overline{Z}_4 + X_2)(\overline{Z}_4 + \overline{B}_0)$$
$$(X_2 + \overline{A}_{-\infty} + \overline{B}_{-\infty})$$
$$(\overline{X}_2 + A_{-\infty})(\overline{X}_2 + B_{-\infty}). \tag{6.4}$$

Thus, $Z_4 = 1$ can be sensitized by assigning $A_{-\infty} = 1$, $B_{-\infty} = 1$, $B_0 = 0$, $X_2 = 1$. This can be obtained by solving SAT of the above characteristic function, Eq. 6.4. The entire waveform is as shown in Figure 6.5.

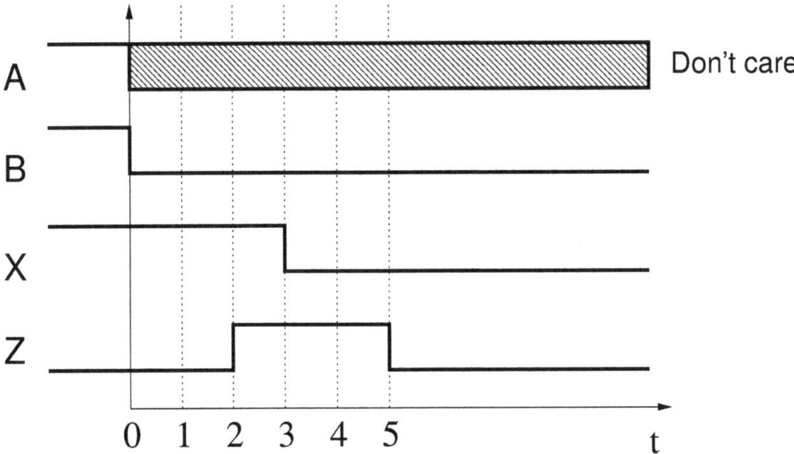

Figure 6.5. Waveform of Example Circuit in Figure 6.8

3.5.2 Translation of Maximum Coupling Effects into an Objective Function

After setting up the circuit characteristic functions, we can formulate an objective function that represent every possible coupling scenario. A rising switching at time t for node X can be translated into $\overline{X}_{t-1}X_t$, where the time resolution is assumed to be 1.0. Similarly, a falling switching at time t for node X can be translated into

$X_{t-1}\overline{X_t}$. Suppose node X and node Z attack node Y (not shown in the figure) at time 3 in Figure 6.4. The maximum coupling function for rising attacking is

$$V_{X,Y}\overline{X}_2 X_3 + V_{Z,Y}\overline{Z}_2 Z_3 - V_{X,Y} X_2 \overline{X}_3 - V_{Z,Y} Z_2 \overline{Z}_3.$$

Because node X has no signal arrival time between time 0 to time 2, we can conclude $X_2 = X_1 = X_0 = X_{-\infty}$ by analyzing all of the possible topological path delays. Similarly, we can say $Z_3 = Z_2$. The objective function is reduced to:

$$V_{X,Y}\overline{X}_0 X_3 - V_{X,Y} X_0 \overline{X}_3.$$

This is equivalent to say only node X can switch at time 3. Meanwhile, X_3 is also constrained by:

$$1 \equiv (X_3 + \overline{A}_0 + \overline{B}_0)(\overline{X}_3 + A_0)(\overline{X}_3 + B_0).$$

For a node i, the maximum rising crosstalk noise is:

$$\sum_j V_{A^j,i}\overline{A}^j_{t-1} A^j_t - V_{A^j,i} A^j_{t-1}\overline{A}^j_t,$$

A_j is the aggressor of node i. Note that a rising and a falling crosstalk noise at the same time caused by two different aggressors can cancel each other in terms of the coupling effects. These A^j_t variables should be constrained to match the circuit behavior as the previous section described.

3.5.3   Boolean Constrained Optimization Problem

The formulation then consists of an objective function which is targeted for the maximum crosstalk noise, and a conjuction of SAT formula to represent the circuit logic behavior considering logic gate delay. This formulation becomes a Boolean Constrained Optimization Problem (BCOP).

The Boolean constrained optimization problem is equivalent to a BCP by transforming the objective function into a cost function, and then minimizing the cost function. A straightforward method to solve this problem is a branch-and-bound method, while many BCP techniques have been proposed [Cou96].

Some heuristics might be possible to reduce the complexity, such as a coarse quantum time, relaxing SAT formulation, or partial variable collapsing.

3.5.4 Discrete Required Time Analysis

In general, the required times under the fixed delay model are not continuous – that is, the possible combinational path delays are finite, resulting in discrete path delays and hence the discrete required times. Therefore, we can analyze these possible discrete required times to reduce the number of timed Boolean variables.

3.5.5 Structural Hashing

In order to reduce the number of timed Boolean variables, this technique tries to find all possible reuses of the timed Boolean variable from the circuit network structure. The simplest reuse is shown in Figure 6.6, where we can reuse $\overline{A}$ for X and replace any occurrence of X by $\overline{A}$.

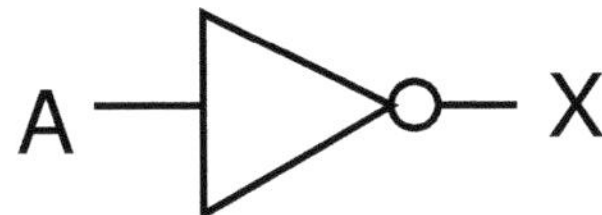

Figure 6.6.   Variable Reuse for an Inverter

When each multi-input gate is represented by, or decomposed into a cube, the localized normal form for gate function representation is established. A hash table can be used to store the output variables with the sorted input list of each gate as a hashing key. The reuse of the circuit in Figure 6.7 can be found by the first reuse $C = \overline{A}$, and then $Y = \overline{X}$. This technique can reduce the number of variables and clauses dramatically, and even reduce the redundant variables at different time points.

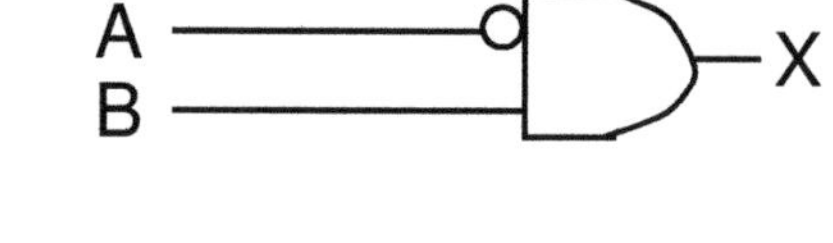

Figure 6.7.   Variable Reuse

3.5.6   Coarse Quantum Time

One way to reduce the variables at the different time points is to assume a very coarse quantum time. Many timed Boolean variables at different time points can be thus collapsed into one. However, this technique leads to a more conservative noise bound.

3.5.7   Boolean Constraint Relaxation

For large circuits, the CNF clauses might be too numerous to solve or satisfy. To relax the functional constraints, it can be just restricted to the relevant sum of product terms, in which we do not attempt to satisfy all of the characteristic functions. The idea is to select a depth of gates to build the characteristic functions, and assume the inputs of the boundary gates without the functional correlation.

4.   Experimental Results

To test our approach, we used the ISCAS85 benchmark circuit set and made some simple assumptions that would emulate accurate layout information.

In actual practice, due to the locality of the layout (i.e. electrical pruning) for each victim net, there are typically only a few aggressors that can cause significant noise. For testing our approach, we emulate this effect by selecting four random aggressor nets and one victim net. These results are shown in Table 6.1. It might be useful to refer back to Section 3.5 to understand the approaches associated with the columns. The first column gives the circuit name from the ISCAS benchmark. The *simple worst* column is the sum of the maximum crosstalk noise from each aggressor. The *zero-delay model* column and the *static noise analysis* column are described in Section 3.3 and 3.4, respectively. The *fixed delay* column is the 2-vector approach described in Section 3.5.

Because we do not have real layout information, we use some electrical parameters, such as C_v, C_x, R_v, t_f, based on a sample of 0.5um 5V static CMOS process, and use Eq. 1.2 to calculate the maximum ΔV_v, the maximum voltage difference shown on the victim net. Arbitrarily, the same electrical parameters are used for each circuit. We assume a condition similar to Fig. 1.8, in which the four aggressor nets possibly make a transition from high to low, and the victim net keeps static high. The numbers shown in Table. 6.1 are the maximum voltage difference on the victim net due to the coupling effect.

For the minimum voltage to be regarded as logic high, it should be within 70% of VDD – that is, ΔV_v should be less than 1.5 Volts in our test case. Therefore, the comparison of the maximum noise bound shows that for C499, C1908 and C2670 circuits, the static noise analysis makes an over-pessimistic prediction, while the fixed delay model does not. The zero-delay model cannot take into account the effect of glitches, and as a result, it under-predicts the maximum noise bound of C1908 and C2670. For most of the cases, the zero-delay model cannot predict tighter maximum noise bound than the static noise analysis, The simple worst case gives nothing informative and is included only as a reference. C6288 could not be completed within a reasonable time.

Circuit	Fixed Delay	Static Noise Analysis	Zero Delay	Simple Worst
C0432	1.91V	2.03V	2.15V	2.15V
C0499	0.84V	1.50V	1.61V	2.15V
C0880	1.74V	1.74V	2.15V	2.15V
C1355	0.76V	0.76V	1.61V	2.15V
C1908	1.40V	1.61V	1.07V	2.15V
C2670	1.19V	1.61V	1.07V	2.15V
C3540	0.54V	0.54V	2.15V	2.15V
C5315	0.54V	0.54V	2.15V	2.15V
C7552	0.39V	0.42V	1.61V	2.15V

Table 6.1.   Comparison of Maximum Noise Bound(Maximum ΔV_v)

5.   Future Work

Our approach aims to find two vectors that maximize the noise under assumptions that are as accurate as possible while being conservative. One way to improve the conservatism of our approach is to consider the effect of cross-coupling on delay degradation, and therefore on timing. It is possible to model the rise waveform on the victim net and compute the maximum delay degradation using Eq. 1.2. However, as more accurate RC-interconnect model is desired for deep submicron technology, the modeling approach similar to [YCGS97, SNEZ97b] should be taken into account.

One area for improvement in the accuracy of our approach is to consider combinational logic blocks in their sequential context. We consider combinational blocks in isolation and presume that the vector pair that we identify is always within the valid sequential state-space of the circuit. In other words, we assume that the vector pair that we identify can be excited in the normal operation of the circuit. This might not be true and thus we could over estimate the noise of the circuit if this is not so. Resolving this issue is more computationally challenging, as it is equivalent to the sequential testing problem, or alternatively, the sequential state space reachability problem, which is currently unsolved for large circuits.

One approach to improving both the accuracy and the conservatism of our method is to incorporate a timing model in which bounded delay intervals, rather than fixed delay values, are used. This approach will be investigated, but currently it appears to make the problem computationally intractable for reasonable sized circuits.

As we were unable to complete our computation on C6288, there is still room for improvement in improving the computational performance of our approach.

For critical path delay degradation, we should consider the extra delay due to the noise interference of the previous input stage. It will result in a fixed-point algorithm as described in Chapter 3 to determine degraded delays on the gates of the critical path, and the objective function should be modified to be the maximum of the critical path delays, which should be computed dynamically.

Our approach is conservative in the sense that we assume signal correlation only within a combinational block, while signals are assumed uncorrelated across sequential gates. Correlation could be possible to cross the sequential gates by the BDD state traversal approach with the timing information such as TBF[LB94]. However, the complexity is even higher than that of the sequential test generation. It is not practical for realistic circuits.

6.   Conservativism Consideration

To make the approaches in this chapter conservative in practice, the delay model needs to take every timing uncertainty into account. For example, a bounded delay model needs to be used. The time slot approach described in Chapter 5 can be used to reduce the complexity, and it has a natural fit by using Boolean variables in the formulation.

7.   Conclusions

The goal of this chapter is to develop an algorithm, software tool, and noise analysis flow that provides a reasonably accurate and conservative approach to the analysis of noise problems that could cause voltage glitches that lead to erroneous switching of dynamic logic or the malfunctioning of analog circuitry. With such a tool available, the time consuming manual work in analyzing potentially noisy signals could be avoided. To achieve improved accuracy our approach finds two vectors that maximize the noise, and we have presented a general scheme for identifying the proper vector pair.

This chapter compares the results obtained by simpler methods including the *zero delay model* in which functional information is incorporated but timing information is neglected, and the *static noise analysis* approach in which temporal information is incorporated but functional information is ignored. Our approach is shown to be strictly more accurate than either of these approaches, while still being computational feasible on industrially sized sub-circuits.

Chapter 7
CONCLUSIONS

We have studied various static noise analysis problems and techniques for DSM designs. The contribution of this work is summarized below:

In Chapter 2, we showed how to use the Miller factors to estimate the extra delay induced by crosstalk effects. This approach uses a decoupled circuit to approximate a coupling circuit. The experimental results showed promising accuracy. A theoretical upper bound of 3X is also found for the opposite direction switching, and a lower bound of -1X for the same direction switching. The conventional 2X factor is shown by experiments not as a bound and can be inaccurate for coupling delay calculation.

In Chapter 3, we developed the mathematical foundation to compute the switching windows. Many numerical properties were formulated and proved. We also studied the effect of using different underlying coupling models, and its associated computation complexity. This work can solve most of the problems regarding switching window convergence.

In Chapter 4, we proposed some event-driven algorithms to speed up the switching window calculation and showed how to efficiently align aggressors to create the worst case delay. We also compared the performance of different scheduling approaches.

In Chapter 5, we further modified the idea of using a continuous switching windows; instead, we proposed a discrete switching window by using time slots. This improves the accuracy and reduces the pessimism in noise analysis. Our experiments showed that up to 90% of potential noise violations suggested by continuous switching windows can be excluded by this approach. Moreover, the size of a time slot can be controlled to trade off accuracy for speed and capacity, which makes this algorithm highly scalable.

In Chapter 6, we presented a functional noise analysis to show how a vector pair can be found to exercise the maximum noise on a given net. We used a SAT formulation to solve the false switching problem, and showed how temporal information should be included to improve accuracy. A Boolean constrained optimization problem was used to find the maximum noise. A similar approach can be used to find if the noise can be propagated to a latch.

The characteristics of DSM processes make crosstalk effects no longer negligible. Due to the high complexity of DSM chips and the high cost of DSM processes, the chip design requires an extensive analysis of possible factors that affect the chip's performance and functionality. The work presented here helps bring a practical crosstalk analysis methodology into the DSM design realm.

References

[AKK+00]   K. Aringaran, F. Klass, C. M. Kim, C. Amir, J. Mitra, E. You, J. Mohd, and S. K. Dong. "Coupling Noise Analysis for VLSI and ULSI Circuits". In *IEEE of 1st International Symposium on Quality Electronic Design*, pages 485–489, Mar. 2000.

[ARP00]   R. Arunachalam, K. Rajagopal, and L. T. Pileggi. "TACO: timing analysis with coupling". In *Proc. of Design Automation Conference*, pages 266–269, 2000.

[BH00]   M. Becer and I. N. Hajj. "An Analytical Model for Delay and Crosstalk Estimation with Application to Decoupling". In *IEEE of 1st International Symposium on Quality Electronic Design*, pages 51–57, Mar. 2000.

[Bry86]   R. E. Bryant. "Graph-based Algorithms for Boolean Function Manipulation". *IEEE Trans. on Computers*, pages 677–691, Aug. 1986.

[CGB97]   W. Chen, S. K. Gupta, and M. A. Breuer. "Analytic Models for Crosstalk Delay and Pulse Analysis Under Non-Ideal Inputs". In *International Test Conference*, pages 809–818, Nov. 1997.

[CK99]   P. Chen and K. Keutzer. "Towards True Crosstalk Noise Analysis". In *Proc. of International Conferences on Computer Aided Design*, pages 132–137, Nov. 1999.

[CKK00a]   P. Chen, D. A. Kirkpatrick, and K. Keutzer. "Miller Factor for Gate-Level Coupling Delay Calculation'. In *Proc. of International Conference on Computer Aided Design*, pages 68–74, 2000.

[CKK00b]   P. Chen, D. A. Kirkpatrick, and K. Keutzer. "Switching Window Computation for Static Timing Analysis in Presence of Crosstalk Noise'. In *Proc. of International Conference on Computer Aided Design*, pages 331–337, 2000.

[CKK00c]   Pinhong Chen, Desmond A. Kirkpatrick, and Kurt Keutzer. "Miller Factor for Gate-Level Coupling Delay Calculation". In *preparation for submission to ICCAD*, 2000.

[CKTK02]   P. Chen, Y. Kukimoto, C.-C. Teng, and K. Keutzer. 'On Convergence of Switching Windows Computation in Presence of Crosstalk Noise". In *Proc. of International Symposium on Physical Design*, pages 84–89, Apr. 2002.

[CMS01]   L. H. Chen and M. Marek-Sadowska. "Aggressor Alignment for Worst-Case Crosstalk Noise". *IEEE Trans. on Computer-Aided Design*, Vol.20:pp.612–621, May 2001.

[Cou96]   O. Coudert. "On Solving Covering Problem". In *Design Automation Conference*, pages 197–202, 1996.

[Dev97]   A. Devgan. "Efficient Coupled Noise Estimation for On-Chip Interconnects". In *Proc. of International Conference on Computer Aided Design*, pages 147–151, 1997.

[DKM93]   S. Devadas, K. Keutzer, and S. Malik. "Computation of floating mode delay in combinational networks: Theory and algorithms". *IEEE Trans. on Computer-Aided Design of Integrated Circuits and Systems*, 12:1913–1923, Dec. 1993.

[DKMW94]   S. Devadas, K. Keutzer, S. Malik, and A. Wang. "Certified Timing Verification and the Transition Delay of a Logic Circuit". *IEEE Trans. on Very Large Scale Integration Systems*, 2:333–342, Sep. 1994.

[DKW92]   S. Devadas, K. Keutzer, and J. White. "Estimation of Power Dissipation in CMOS Combinational Circuit Using Boolean Function Manipulation". *IEEE Trans. on Computer-Aided Design of Integrated Circuits and Systems*, 11:373–383, Mar. 1992.

[DMP96]   F. Dartu, N. Menezes, and L. T. Pileggi. "Performance computation for precharacterized CMOS gates with RC-loads". *IEEE Trans. on Computer-Aided Design of Integrated Circuits and Systems*, 15:544–553, May 1996.

[DP97]   F. Dartu and L. T. Pileggi. "Calculating Worst-Case Gate Delays Due to Dominant Capacitance Coupling". In *Proc. of 34th ACM/IEEE Design Automation Conference*, Jun. 1997.

[ea97]   S. Tasiran et. al. "Comuting Delay with Coupling with Timed Automata". In *Tau 97 : International Workshop ion Timing Issues*, pages 232–242, 1997.

[EMU96]   G. Engeln-M. and F. Uhlig. *"Numerical Algorithms with C"*. Springer-Verlag, 1996.

[FFP+00]   B. Franzini, C. Forzan, D. Pandini, P. Scandolara, and A. D. Fabbro. "Crosstalk Aware Static Timing Analysis:a Two Step Approach". In *IEEE of 1st International Symposium on Quality Electronic Design*, pages 499–503, Mar. 2000.

[GRP98]   P. D. Gross, R. Arunachalam K. Rajagopal, and L. T. Pileggi. "Determination of Worst-Case Aggressor Alignment for Delay Calculation". In *Proc. of International Conferences on Computer Aided Design*, pages 212–219, Nov. 1998.

[Kir97]   D. A. Kirkpatrick. *"The Implication of Deep Sub-micron Technology on the Design of High Performance Digital VLSI System"*. PhD thesis, CAD Group Ph.D. Dissertation, U.C. Berkeley, 1997.

[KMS00]   A. B. Kahng, S. Muddu, and E. Sarto. "On Switch Factor Based Analysis of Coupled RC Interconnects". In *Proc. of Design Automation Conference*, pages 79–84, 2000.

[KMV99]   A. B. Kahng, S. Muddu, and D. Vidhani. "Noise and Delay Estimation for Coupled RC Interconnects". In *IEEE AISC/SoC*, Mar. 1999.

[Lar92]   T. Larrabee. "Test Pattern Generation Using Boolean Satisfiability". *IEEE Trans. on Computer-Aided Design of Integrated Circuits and Systems*, 11:4–15, Jan. 1992.

[LB94]   W. K.C. Lam and R. K. Brayton. *"Timed Boolean Functions"*. Kluwer Academic Publishers, 1994.

[LBB+00]   R. Levy, D. Blaauw, G. Braca, A. Dasgupta, A. Grinshpon, C. Oh, B. Orshav, S. Sirichotiyakul, and V. Zolotov. "Clarinet: A noise analysis tool for deep submicron design". In *Proc. of Design Automation Conference*, pages 233–238, 2000.

[QPP94]   J. Qian, S. Pullela, and L. T. Pillage. "Modeling the "effective capacitance" of the RC interconnect". *IEEE Trans. on Computer-Aided Design of Integrated Circuits and Systems*, Vol.13:pp.1526–1535, Dec. 1994.

[RIZK94]   A. Rubio, N. Itazaki, X. Zu, and K. Kinoshita. "An Approach to the Analysis and Detection of Crosstalk Faults in Digital VLSI Circuits". *IEEE Trans. on Computer-Aided Design*, 13:387–394, Mar. 1994.

[Sap99]   S. S. Sapatnekar. "On the Chicken-and-Egg Problem of Determining the Effect of Crosstalk on Delay in Integrated Circuits". In *Proc. IEEE of 8th Topical Meeting on Electrical Performance of Eletronic Package*, pages 245–248, 1999.

[Sap00]   S. S. Sapatnekar. "A timing model incorporating the effect of crosstalk on delay and its application to optimal channel routing".

IEEE Trans. on Computer-Aided Design, Vol.19:pp.550–559, May 2000.

[Sas93]   T. Sasao(ed). *"Logic Synthesis and Optimization, Ch.8: Delay Models and Exact Timing Analysis"*. Kluwer Academic Publishers, 1993.

[SBSV96]   P. Stephan, R. K. Brayton, and A. L. Sangiovanni-Vicentelli. "Combinational Test Generation Using Satisfiability". *IEEE Trans. on Computer-Aided Design*, 19:4–15, Sep. 1996.

[She98a]   K. L. Shepard. "Design Methodologies for Noise in Digital Integrated Circuits". In *Design Automation Conference*, pages 94–99, 1998.

[She98b]   K. L. Shepard. "Design Methodologies for Noise in Digital Integrated Circuits". In *Proc. of Design Automation Conference*, pages 94–99, 1998.

[SNEZ97a]   K. L. Shepard, V. Narayanan, P. C. Elmendor, and Gutuan Zheng. "Global Harmony: Coupled Noise Analysis for Full-Chip RC Interconnect Network". In *Proc. of International Conference on Computer Aided Design*, pages 139–146, 1997.

[SNEZ97b]   K. L. Shepard, V. Narayanan, P.C. Elmendor, and Gutuan Zheng. "Global Harmony: Coupled Noise Analysis for Full-Chip RC Interconnect Network". In *International Conference on Computer Aided Design*, pages 139–146, 1997.

[SNEZ97c]   K. L. Shepard, V. Narayanan, P.C. Elmendor, and Gutuan Zheng. "Global Harmony: Coupled Noise Analysis for Full-Chip RC Interconnect Networks". In *Proc. of International Conference on Computer Aided Design*, pages 139–146, 1997.

[TCE00]   P. F. Tehrani, S. W. Chyou, and U. Ekambaram. "Deep Sub-Micron Static Timing Analysis in Presence of Crosstalk". In *IEEE of 1st International Symposium on Quality Electronic Design*, pages 505–512, Mar. 2000.

[XCMS00]   T. Xiao, C. W. Chang, and M. Marek-Sadowska. "Efficient static timing analysis in presence of crosstalk". In *Proc. of IEEE International ASIC/SOC Conference*, pages 335–339, 2000.

[XMS00a]   T. Xiao and M. Marek-Sadowska. "Efficient Delay Calculation in Presence of Crosstalk". In *IEEE of 1st International Symposium on Quality Electronic Design*, pages 491–497, Mar. 2000.

[XMS00b]   T. Xiao and M. Marek-Sadowska. "Worst delay estimation in crosstalk aware static timing analysis". In *Proc. of IEEE International Conference on Computer Design*, pages 115–120, 2000.

[YCGS97]   G. Yee, R. Chandra, V. Ganesan, and C. Sechen. "Wire Delay in the Presence of Crosstalk". In *IEEE/ACM International Workshop on Timing Issues in the Specification and Synthesis of Digital Systems*, pages 170–175, 1997.

[ZSN01]   H. Zhou, N. Shenoy, and W. Nicholls. "Timing Analysis with Crosstalk as Fixpoints on Complete Lattice". In *Proc. of Design Automation Conference*, pages 714–719, 2001.